I0881838

ENVIRONMENT AND BELIEF SYSTEMS

Part of the series Key Concepts in Indigenous Studies, this book focuses on the concepts that recur in any discussion of nature, culture and society among the indigenous.

The book, the first in a five-volume series, deals with the two crucial concepts of environment and belief systems of indigenous peoples from all the continents of the world. With contributions from renowned scholars, activists and experts from around the globe, it presents a salient picture of the environments of indigenous peoples and discusses the essential features of their belief systems. It explores indigenous perspectives related to religion, ritual and cultural practice, art and design, and natural resources, as well as climate change impacts among such communities in Latin and North America, Oceania (Australia, New Zealand and the South Pacific Islands), India, Brazil, Southeast Asia and Africa.

Bringing together academic insights and experiences from the ground, this unique book's wide coverage will serve as a comprehensive guide for students, teachers and scholars of indigenous studies. It will be essential reading for those in anthropology, social anthropology, sociology and social exclusion studies, religion and theology, and cultural studies, as well as activists working with indigenous communities.

G. N. Devy is Honorary Professor, Centre for Multidisciplinary Development Research, Dharwad, India, and Chairman, People's Linguistic Survey of India. An award-winning writer and cultural activist, he is known for his 50-volume language survey. He is Founder Director of the Adivasi Academy at Tejgadh in Gujarat, India, and was formerly Professor of English at M. S. University of Baroda. He is the recipient of the Sahitya Akademi Award, Linguapax Prize, Prince Claus Award and Padma Shri. With several books in English, Marathi and Gujarati, he has co-edited (with Geoffrey V. Davis and K. K. Chakravarty) *Narrating Nomadism: Tales of Recovery and Resistance* (2012); *Knowing Differently: The Challenge of the Indigenous* (2013); *Performing Identities: Celebrating Indigeneity in the Arts* (2014); and *The Language Loss of the Indigenous* (2016), published by Routledge.

Key Concepts in Indigenous Studies

Series Editor: G. N. Devy

Honorary Professor, Centre for Multidisciplinary Development Research, Dharwad, India, and Chairman, People's Linguistic Survey of India

This series of volumes offers the most systematic and foundational literature available to date for use by undergraduate and postgraduate students of indigenous studies. It brings together essays by experts from across the globe on concepts forming the bedrock of this rapidly growing field in five focused volumes: *Environment and Belief Systems* (Vol. 1); *Gender and Rights* (Vol. 2); *Indigeneity and Nation* (Vol. 3); *Orality and Language* (Vol. 4); and *Performance and Knowledge* (Vol. 5). These contain short, informative and easily accessible essays on the perspectives of indigenous communities from all continents of the world. The essays are written specifically for an international audience. They thus allow drawing of transnational and cross-cultural parallels, and form useful material as textbooks as well as texts for general readership. Introducing a new orientation to traditional anthropology with comprehensive and in-depth studies, the volumes foreground knowledge traditions and praxis of indigenous communities.

Environment and Belief Systems
Edited by G. N. Devy

Gender and Rights
Edited by G. N. Devy

Indigeneity and Nation
Edited by G. N. Devy

Orality and Language
Edited by G. N. Devy

Performance and Knowledge
Edited by G. N. Devy

For more information about this series, please visit: www.routledge.com/Key-Concepts-in-Indigenous-Studies/book-series/KCIS

ENVIRONMENT AND BELIEF SYSTEMS

Edited by G. N. Devy

Routledge
Taylor & Francis Group
LONDON AND NEW YORK

First published 2021
by Routledge
2 Park Square, Milton Park, Abingdon, Oxon OX14 4RN

and by Routledge
52 Vanderbilt Avenue, New York, NY 10017

Routledge is an imprint of the Taylor & Francis Group, an informa business

British Library Cataloguing-in-Publication Data
A catalogue record for this book is available from the British Library

Library of Congress Cataloging-in-Publication Data
A catalog record for this book has been requested

ISBN: 978-0-367-24517-7 (hbk)
ISBN: 978-0-367-41018-6 (pbk)
ISBN: 978-0-367-81427-4 (ebk)

Typeset in Bembo
by Apex CoVantage, LLC

This volume and four others in this Key Concepts in Indigenous Studies series are dedicated to the memory of Geoffrey Davis, a life-long friend of the indigenous and the marginalized, formerly Professor of English at RWTH, Aachen, Germany, and Chairperson, Association of Commonwealth Literature and Language Studies, whom I knew as a friend for over three decades, with whom I collaborated on several intellectual projects and who would have been co-editor for this series but for his unexpected demise before the work was completed.

CONTENTS

FIGURES

CONTRIBUTORS

Penny Allan is Professor of Landscape Architecture at the University of Technology Sydney, Australia. She is a co-founder of the design research practice OUTPOST. Her three most recent design research projects *MOVED to Design, Earthquake Cities of the Pacific Rim*, and *Rae ki te Rae* deal with the relationship between environment, culture, resilience and design, and have all received national awards. Her latest research looks at cultural resilience and climate change migration in the Pacific.

Tracey Benson is an interdisciplinary artist, researcher and creative producer based in Australia. She has over 15 years of experience as an executive in the Australian Public Service managing ICT, environment and behaviour change programs. She is currently a consultant to the public sector and an adjunct professional associate at the Institute of Applied Ecology at the University of Canberra. Recent publications include chapters in *Mobile Media Practices, Presence and Politics* and *Locating Emerging Media*.

Martin Bryant is a landscape architect, architect and urban designer with more than three decades of experience in private practice and academia. He is Head, Architecture School at University of Technology Sydney, Australia. His multi-award-winning design projects are recognized for environmental benefits, cultural sensitivity and urban design. His globally significant research led to his authorship of an urban ecology and resilience policy paper for the 2016 United Nations Habitat III conference in Quito. He continues to research in designing for resilience.

James L. Cox is Professor Emeritus, University of Edinburgh, United Kingdom.

Charles Dawson directs an interdisciplinary New Zealand study programme for the Minnesota-based social- and environmental justice consortium HECUA.

His recent publications attend to bicultural water issues (in *Ecocriticism of the Global South*) and climate and culture for Cambridge's *Global History of Literature and the Environment*.

Seth Garfield is Assistant Professor of History, University of Texas at Austin, Texas, USA.

Oscar Guardiola-Rivera is Senior Lecturer in Law, Birkbeck College, University of London, United Kingdom.

Lee Joachim is Yorta Elder, Researcher and CEO of the Rumbalara Aboriginal Co-operative in Shepparton, Australia. He has contributed to many publications addressing climate change, sustainability and indigenous knowledge.

Penehuro Fatu Lefale is Director of LeA International (New Zealand), a company that specializes in climate science, climate policy, international development, and national meteorological and hydrological services (NMHS) management systems. He is also an external affiliate researcher with the Joint Centre for Disaster Research (JCDR), Massey University, Wellington, New Zealand.

Deborah McGregor is Professor at Osgoode Hall Law School Faculty and Canada Research Chair in Indigenous Environmental Justice, Faculty of Environmental Studies, York University, Toronto, Canada.

Archana Prasad is Professor and Chair, Centre for Informal Sector and Labour Studies, Jawaharlal Nehru University, Delhi, India.

Huhana Smith (Ngāti Tukorehe, Ngāti Raukawa ki Te Tonga) is an artist and academic with wide-ranging experience in Māori visual art and museum practice, exhibition planning and implementation, indigenous knowledge and science research. She is currently Head of Whiti o Rehua, School of Art, Toirauwhārangi, College of Creative Arts at Massey University, Wellington, New Zealand. She has co-led another active participatory, multi-disciplinary, collaborative and kaupapa Māori research project called "Risk Management Planning for Climate Change Impacts on Māori Coastal Ecosystems and Economies."

Tamasailau Suaalii-Sauni is Associate Professor in the Criminology Programme, School of Social Sciences, Faculty of Arts, University of Auckland, New Zealand, with wide-ranging interests in law and custom, indigenous jurisprudence and criminology in the Pacific and Pacific indigenous research methodologies.

Michael Wessels was Associate Professor, Department of English, University of the Western Cape, South Africa.

PREFACE

The volumes in this series have long been in making. The idea came up in 2011 in a conversation between Prof. Geoffrey Davis and me. The two of us had by then worked on six anthologies related to Indigenous Studies to which scholars from all continents had contributed. Two of these are published by Orient BlackSwan (*Indigeniety*, 2009 and *Voice and Memory*, 2011) and four by Routledge between 2012 and 2016 (*Narrating Nomadism, Knowing Differently, Performing Identities* and *The Language Loss of the Indigenous*). However, we felt that we need to do more, a lot more, in order to firmly establish this newly emerging field. Shashank Sinha and Shoma Choudhury of Routledge showed a keen interest in our proposal. Enthused by the idea of bringing out a set of volumes dealing with some of the definitive themes of the field, and assured by the possibility of publication of the volumes, we started our work. Of course, it was not an entirely easy going for us. The challenges were many and the scale in which we wanted to cast the volumes was not easy to handle. Despite the difficulties and setbacks expected in such an intellectual venture, we kept up. Most of the editorial was completed by early 2018. As we were getting ready to send the typescripts, alas, Prof. Geoffrey Davis died in a short internment in an Aachen hospital. His last mail came to me a day before he was to be admitted. The loss was a big blow to me. His friends and colleagues spread over all continents deeply mourned his death. For me, the most civilized way of mourning is to ensure that the volumes to which he had contributed so much care and toil get published. Who was Geoffrey Davis and why was he interested in the indigenous? Perhaps the best way for me to explain this is to repeat here the response I sent to two questions from Prof. Janet Wilson (hereafter JW) of the Southampton University.

JW: What were the points of synergy (ideological, intellectual, political activist) that brought you and Geoff together, and when and how did this happen, i.e., what were the particular contexts/motivations? I remember I think Geoff had just retired and was possibly looking for a new project? And might have been inspired through his involvement with ACLALS.

DEVY: I think I met him the first time in 1984 at the EACLALS conference at Sitges, Spain. During the 1980s, I was a "regular" at the EACLALS since India did not have an active Commonwealth Literature culture as yet. But, my memory of that meeting is not very clear. In 1988, Geoff had convened a conference at Aachen, Germany, where Geoff spent most of his academic life. I was invited to it for a plenary. This experience left me impressed by his organisational ability. In between, we had met at other places too, Austria, Hungary, Singapore. But all these meetings were casual; and I do not recall any memorable conversation having taken place between us during these conferences. During the 1990s, Geoff hosted a conference on Literature and Activism. I left my professorship at Baroda, India, in 1996. Geoff had heard about this move from friends. He asked me to lecture at the conference. I did. It was during this conference I noticed that he was deeply respectful of activism, that his empathy for the dispossessed was genuinely deep. I also noticed that he was extremely wary of using clichéd and fashionable jargon. The impression these qualities made on me was strong. A few years later, he was to attend the ACLALS Triennial in Hyderabad, India. He wrote to me asking if he could visit me after the Hyderabad conference. He knew that I had stopped attending academic conferences and there was no chance of our meeting in Hyderabad. So, I invited him to Baroda, a 1500 km north of Hyderabad.

I am not sure if he enjoyed his visit to Baroda. On the day he was to arrive, for reasons difficult for to me know, I forgot about his arrival altogether. I was to meet him at the airport. Baroda in those days was a very small airport and every day only three or four flights arrived there. And overseas visitors were not a common sight. Geoff waited there till almost the last co-traveller had left the meeting area. The last one to leave happened to be an architect named Karan Grover, who is a living legend in the field of architecture. Grover asked Geoff if he was expecting anyone. Geoff mentioned my name. This worked. Karan Grover and I had been friends for decades and Geoff was made to feel welcome on my behalf, brought to his lodgings and, the forgetting and forgiving over, we met over dinner. The next day, I drove him in my car to the location of the Adivasi Academy (the Tribal University) that I was trying to establish in those years. This location was 90 km east of Baroda. On the way, I talked with passion about all my plans, my dreams. He listened. He spent another day in Baroda meeting Karan and enjoyed the famous Grover wine. I was busy in my work with the tribal academy. The next morning, I drove Geoff to the airport as he was leaving for Bombay and then to Aachen. At the airport, he asked me if I could have him visit the Adivasi Academy again for a longer time, perhaps a week or so. I said, "Why do you not stay for a semester?" He was a bit puzzled by my offer, made in such a casual manner. So, I added, "Be a Fellow with us." He took that offer and returned to Baroda the following year, but for a short time. I think it was after two more brief trips that he agreed to spend six months in Baroda.

I must explain that the Adivasi Academy is not like a university. It is a community workstation at best, with really very minimum facilities that makes for most of us what we call "civilization." The "Fellowship" had no set rules. They were made looking at the individual's ability and desire to contribute what one had promised to contribute. The "projects" ranged from writing a book or an article, teaching music or language to children, keeping the library or museum in good order, tending a piece of agricultural land, setting up a community micro-credit group or just documenting any of these activities. When Geoff became a Fellow of the Adivasi Academy, there were three others, Brian and Eileen Coates from Limerick, Ireland and Lachman Khubchandani, a Linguist from Pune, India. Eileen had agreed to help us with the Museum and Lachman was to write a book in linguistics. I was more ambitious with Geoff. I said to him, "If you do not mind, please do nothing, only watch what goes on here and when it pleases you discuss ideas with me." He agreed. The facilities given to the Fellows included housing in Baroda and meals when they visited the Academy, 90 km away from Baroda. All my meetings with tribals were transacted in their languages. English words were rarely heard. Only occasionally, some visitors helped Geoff with English interpretations. Geoff, I must say, braved all of this discomfort without a murmur. The impression I had formed about his deep empathy for the dispossessed became firmer. In the fifth month of his stay, I sent a word to him asking if he was available for a serious conversation. He obliged. We met in my Baroda office – the Bhasha Centre – at 2 PM. I asked him if he would join me in imagining an international "non-conference" for looking at the world through the perspective of the indigenous. He said, "I cannot promise, but I will try." Our conversation continued for several hours and, probably, both of us had a reasonably good idea of what must be avoided in making our idea of a conference completely rooted to the ground. I proposed the name "chotro" (a shared platform); he consented to it with great enthusiasm. Next morning, I found him at Bhasha. He had a "Call for Chotro" ready with him. I made several calls to various offices and individuals in Delhi to finalize the material arrangements for the First Chotro. That afternoon, Geoff sat at the computer and sent out close to 150 emails. Before he left Baroda, we were fully involved in putting together the unusual conference. He made one visit to India before the conference was held in Delhi in January 2008. We met in Delhi. I had to combine some of my other works with the work related to Chotro. One of these involved a visit to the prime minister's office. He was a bit shocked when I told him that after sorting out the conference-related arrangements for stay and local transportation, I would be going to the PM's office and he was welcome to join me. Years later, I heard him narrating this anecdote to friends over a glass of wine. The Delhi Chotro was the first one. We put together several more in subsequent years and worked on the conference volumes, meeting in several countries. Geoff became a frequent visitor to India to Baroda, and also to my home and family. I am not aware if we shared an ideology. In a way, all of us in the field of literature have varying degrees of a progressive outlook on life and society. But, what clicked between Geoff and me is something else, and that is his immense patience with my and his ability to cope with surprises and shocks, which could not be avoided considering my involvement in several social causes. John Keats, speaking of William Shakespeare's "genius" used the term "negative capability" – the ability to live amidst uncertainties. The mutual recognition of this negative capability brought us together for undertaking an unconventional kind of work – serious, though not strictly academic.

JW: What roles/or positions did Geoff take as collaborator e.g. in co-organising Chotro and in working with the Adivasis/Bhasha more generally?

DEVY: When we thought of creating the Chotro non-conferences, we had no funding support. We had no sponsors, no funds for international travel. Bhasha Centre was not a full-scale "institution" till then. Besides, "Indigenous Studies" was not any accepted field of academic work. We were not sure if any self-respecting publisher would accept to publish the proceedings. Therefore, in all of these matters, we shared responsibility. But, generally speaking, he dealt with the overseas participants and I handled the Indian issues, material and academic. I accepted to identify publishers, negotiate with them, do the necessary correspondence, Geoff focused on copy editing of the texts. But, this division of work was not sanctimonious. Either of us was free to cross over and even required to do so looking at each other's convenience. Never forget that Geoff had his other major obligations and academic projects, and I had mine. We had no desire to claim credit for the work we were doing. It was born out of our desire to create a legitimate space for the voice of the indigenous.

I hope my response to Janet Wilson will have made it clear why I enjoyed working with Geoffrey Davis on so many intellectual projects. In India's intellectual history, there have been glorious examples of intellectual collaboration between Indian thinkers and scholars, and writers and scholars from other countries. W.B. Yeats and Purohit Swamy, Yeats and Tagore, and Tolstoy and Gandhi developed their ideas through such collaborations. In our time, the rising tide of the Right political parties and a narrow idea of nationalism is gaining a greater currency that makes such collaborations difficult to carry through. I am pleased that this series of volumes is seeing the light of the day, bringing my work together with Geoffrey Davis to a successful conclusion.

ACKNOWLEDGEMENTS

The initial idea of this volume and the series to which it belongs came up in 2012. Since then, Prof. Geoffrey Davis, who was to be the co-editor, corresponded with several scholars from the field of commonwealth literature from other academic disciplines. These scholars from various disciplines and several continents gave their advice and suggestions for identifying scholars to be involved in the project. They are too numerous to be mentioned individually. I would like to record my gratitude to them. The scholars and activists who consented to contribute, and a majority of them who kept their promise, made putting together the volumes possible. Their participation in the most tangible way calls for my thanks. Several organizations and institutions offered Prof. Geoffrey Davis and me opportunities of meeting and taking forward our plans for these volumes. They include the Association for Commonwealth Literature and Language Studies, for a conference in Cyprus; Bhasha Research and Publication centre, Baroda, for various events through these years; the Kiel Voche organized by Kiel City Council in Germany; the German Academy for convening a conference at Hamburg; Aide et Action, for convening a meeting in Geneva; and several Indian colleges and universities, for creating spaces on the sidelines of conferences – I thank all of them. I wonder if the project would have moved forward at all without these meetings.

I would like to thank Ingrid, Prof. Davis's wife, for ungrudgingly encouraging him to spend his funds and time on travels to India to work on this project. Surekha, my life partner, has most generously supported the project throughout its years of slow progress by providing ideas, hospitality and courage. I cannot thank her enough.

The publication of these volumes would not have been at all possible had it not been for the abiding friendship and support of Dr. Shashank Shekhar Sinha, Publishing Director, and his inspiring colleague Shoma Choudhury at Routledge. I carry in my heart the comfort drawn from their genuine friendship.

INTRODUCTION

G. N. Devy

Identity, environment, language, gender, belief systems, performance traditions and rights are some of the more central issues relating to the struggles and the survival of the indigenous all over the world. Their local features vary from community to community and from country to country. However, the general narrative is fairly common. Quintessentially, this narrative refers to a colonial experience that hammered a break in the long-standing traditions of the indigenous, and yet they kept close to their traditions and close also to nature, losing in the process the control over natural resources, land, rivers, and forests and coming in clash with a radically different framework of justice, ethics and spirituality. For the indigenous, invariably, there are two points in time marking their emergence: one that is traced back to a mythological time enshrined in their collective memory and expressed in their community's "story of origin," and the other that is synchronous with a Columbus or a Vasco de Gama setting foot on the land that was once their dominion. It is true that no established research or theory in archaeology, anthropology, genetics, cultural geography, historical linguistics, agriculture and forestry goes to show that all or any of the indigenous people have been inhabitants of the land where they were when colonialism was inaugurated, and that a very small portion of them have been associated with their present habitat since the time the homo sapiens have inhabited the Earth. There were migrations from place to place and from continent to continent during the pre-historic times as well. Yet, despite pre-historic migrations, it is true that the indigenous communities have been associated with their habitats for a considerably long time. The European colonial quest, the territorial and cultural invasion associated with it, and the interference of alien political, ecological and theological paradigms brought a threat to the traditions that the indigenous had developed. The absence of desire on their part to accept the new paradigms and to internalize them made them stand out, be marked as "others," interpreted as "primitive" and represented as "indigenous." It is common sense that term "indigenous" as a part of a binary can have meaning only when there are other terms such as "alien," "outsiders," "non-native" and "colonialists." The one without the other would cease to have the meaning that it now has.

Though the census exercises in different countries do not use a uniform framework, methodology and orientation, the data available through the census carried out by different nations shows that approximately 370 million of the world population is indigenous. The communities identified as indigenous on the basis of their location, uniqueness of tradition, social structures and community law number close to 5,000 and are spread over 90 countries (World Bank, Indigneous People). Several different terms are used for describing them in different continents: "aborigine" "janjati," "indigenous," "First Nations," "natives," "Indian" and "tribe." In most countries, the identification and listing of such communities is by no means complete and has remained, over the last seven decades, since the United Nations Organization was set up, an unfinished process. Despite the inadequacy in the world's knowledge about the indigenous, it is clear that their existence, environment, cultural ethos, lifestyles and values have been under a relentless assault by the practices, culture and values of the rest of the world. In recognition of the threat to indigenous cultures and knowledge systems, to their land and environment, to their languages, livelihood and law, the United Nations came out with a Declaration on the Rights of Indigenous People, accepted by the UN General Assembly in September 2007. The following three Articles of the Declaration address the most fundamental issues involved in the genocidal threat to their survival and their unique cultures:

> **Article 25**
>
> Indigenous peoples have the right to maintain and strengthen their distinctive spiritual relationship with their traditionally owned or otherwise occupied and used lands, territories, waters and coastal seas and other resources and to uphold their responsibilities to future generations in this regard.
>
> **Article 26**
>
> 1 Indigenous peoples have the right to the lands, territories and resources which they have traditionally owned, occupied or otherwise used or acquired.
> 2 Indigenous peoples have the right to own, use, develop and control the lands, territories and resources that they possess by reason of traditional ownership or other traditional occupation or use, as well as those which they have otherwise acquired.
> 3 States shall give legal recognition and protection to these lands, territories and resources. Such recognition shall be conducted with due

> respect to the customs, traditions and land tenure systems of the indigenous peoples concerned.

Article 27

> States shall establish and implement, in conjunction with indigenous peoples concerned, a fair, independent, impartial, open and transparent process, giving due recognition to indigenous peoples' laws, traditions, customs and land tenure systems, to recognize and adjudicate the rights of indigenous peoples pertaining to their lands, territories and resources, including those which were traditionally owned or otherwise occupied or used. Indigenous peoples shall have the right to participate in this process.

Given that the population of the indigenous is less than five percent of the world's total population, and also that they are sharply divided in terms of tribe and community within a given country, every indigenous community exists as a miniscule minority within its political nation. To be indigenous is, in our time, to be severely marginalized in economy, politics, institutionalized knowledge and institutionalized religion. The space for the indigenous is rapidly shrinking. This year, 2019, has been declared by UNESCO as the Year of the Indigenous Languages. There have been official celebrations and academic conference to "celebrate" the year. However, it is a fact that several thousand of the languages still kept alive by the communities are close to extinction. A comprehensive survey of languages that I had conducted of the 780 living languages in India in 2010 showed that nearly 300 languages, all of these spoken by the indigenous peoples, may disappear in the next few decades. In India, the national government passed a law in 2008 requiring the land ownership of tribal communities to be returned to them. However, nearly half of the claims have yet to be settled and the Supreme Court of India has already asked the families whose land title claims not been accepted to evacuate them. This is the situation not only in India. It is so in Thailand, where the tribals in the Chang Mai area have been fighting a bitter battle for land ownership. In Mexico, they are struggling to keep their languages alive; in Australia, despite the best efforts by government, the status of the aborigines in professions and educational institutions remains far from what was visualized; in New Zealand, they had been facing social discrimination and continue to do so; in Africa, their plight still deserves the description "a genocidal neglect." Despite legal provisions aimed at safeguarding the communities and their cultures, they are diminishing and suffering an undeserving obsolescence in the world that is considering when to officially announce that the Anthropocene, the epoch of unprecedented interference with nature fundamentally altering the Earth, has commenced.

On 29 May 2019, I received a press release in my email inbox from ESRC, an international network for economic and social rights network. It referred to an ongoing struggle of the Ogoni people in Nigeria whose land is slotted to be used for oil drilling. It claims that

> The said oil operation, with military cover, is billed to commence in the coming months, as parties involved have been instructed on their respective roles. Each of the security agencies listed in the plan has been tasked with specific roles and responsibilities, particularly, in tackling voices of dissent in the Ogoni community.

It rued the fact that

> rather than going through the proper and legitimate means and process, the government chooses to ignore the people of Ogoni and prefers to engage and impose an oil prospecting company on them, even with an intent to intimidate, suppress and kill the people more with the use of heavily armed security forces.

The press release concluded by reminding the readers that the Ogoni people have never opposed government activity in the Ogoniland, but have insisted on "the issues of benefits-sharing, community participation and the proper environmental management of the Ogoni ecosystem, including legacy issues arising from the over four decades of reckless oil operations in the land."

For nearly a quarter century, since the internet became a means of communication, I have been receiving such statements in my mailbox from the indigenous communities or organizations working on their behalf. They bring a narrative in which the characters change but the plot is predictably the same. It involves a rather helpless indigenous community trying to establish that the natural resource involved is its legacy and a corporate body – supported by the government and aided by the police, military or private security troops – is denying the community's rights. Normally, the conclusion of this story is in the slow dissemination of the community voice and in a gradual success of the state in imposing its will on the community. Rarely do the indigenous succeed in getting their voice heard and respected. From Paraguay to Malaysia and from Canada to Australia – west, east, north, south – the story of conquest of the natural resources of the indigenous peoples by the "civilized" – read exploitative economies and industrial technologies – has been commonplace for the last few decades.

An unending environmental degradation of habitats of the indigenous has been the norm, not an exception, implied in the massive movement of capital across countries. Exploitation of the natural resources has left the traditional habitats of the indigenous people devastated. This situation is not peculiar to any single country or continent; it is the general condition of the indigenous all over the world. In a photo essay in the *National Geographic* of 28 April 2017, Michael

McBride presents a set of moving photographs of various sites of devastation in North America and records:

> The first resource extracted from indigenous Americans was gold. When the (extinct) Arawak ran out of gold to bring Christopher Columbus, he turned to the people themselves as the resource, garnering the majority of his profits by selling them as slaves. This set the tone for dealings with indigenous cultures by the captains of industry who were to be the real conquerors of the "New World". There are countless cases throughout history of tribal peoples being dislocated, abused, and even killed by resource industries and the government policies that support them. Homelands have been lain to waste. Subsistence cultures and the societies tied to them have disappeared along with traditional food sources and even potable water. For centuries indigenous people have paid the major share of the cost of civilization's insatiable appetite for land and resources.[1]

These observations and the visuals are true of the environs of practically every indigenous community in any part of the world. If there is anything more seriously devastated than the land, land resources, forests, rivers and hills of the indigenous, it is, probably, the belief-systems of these communities.

Not confined to sacred books, monuments for divinity and theological bureaucracies, the worship practices of the indigenous have, probably, not even been understood by the "civilized world." It is a normal practice in formal theology and philosophy to categorize the belief practices of the indigenous as "nature worship." Some of the terms used to describe the concept are "druidism," "naturism" and "primitive faith." If one were to ask indigenous if these terms are effective for one's concept of the sacred, the answer is likely to be an amused "hardly so." The first indigenous community with whom I started working was that of the Rathwa Bhils in western India. The Rathwas have houses adorned with wall-size painting of many mythological and historical events. Taken together, they call these paintings "Babo Pithoro." They do not look at this set of images, painted collectively in an elaborate ritual, as a "thing," a representation of "something" or an "isness." They consider the process of making this imagery as their god, the very highest one, with a series of other minor ones painted within the fresco. Many other indigenous communities in India with whom I subsequently worked told me how their beliefs are shaped. The diversity among their metaphysical thought was far beyond what I had initially imagined. Their relationship with nature is respectful, almost prayerful; but that does not mean that it is nature alone that forms the main substance of their religion or belief. I have come across communities which maintain that the highest form of prayer is possible only when one is possessed; and in that state, the possessed body becomes the object of worship by the possessed mind. Institutionalized and textual religions do not allow us the space to create any adequate analogy to describe most of the belief systems of the indigenous.

James L. Cox's discussion of indigenous religions in Oceania and the southern Pacific (Chapter 5) stretching from Melanesia to the eastern Polynesian Islands ably defends Platvoet's thesis that

> The belief systems and social structures within indigenous societies revolve around ancestors and hence the religions of such societies are localised as opposed to universal. Indigenous societies also relate exclusively to a particular geographical place fixed in the tradition by their quasi-legendary myths. Unlike religions with universal cosmologies, *one can never be converted into an indigenous religion, since ancestors belong only to precise kinship groups that operate within restricted and delineated spatial locations.*
>
> *(Platvoet 1992: 21–22; Cox 20)*

It is not generally recognized by colonialism scholars that the earliest clash between the indigenous and their colonizer-adversaries was not so much related to territory and resources but about belief. European nations that went out in the world to colonize civilizations in other continents had put forward the gospel as their frontier weapon. As Oscar Guardiola-Rivera's nuanced essay in this volume (Chapter 7) points out,

> From the outset, Amerindians and their allies questioned the centrality and totalising character of belief, casting suspicion on the idea of society as a calculated totality founded on the obsessive policing of a hard core and its limits, which uses the exterior as an inverted mirror for self-identity. Put simply, Amerindians were suspicious of their Greek-Christian fellows bearing the gift of totalising belief or "true faith", in a way like the Brazilian gentleman in Costas' story.
>
> *(07: 61–71)*

A community's environment and its belief patterns are deeply interconnected. The ideas of what is sacred and deserving of worship are shaped by the environment defining the life conditions of a given community. Similarly, the manner in which a community perceives the elements in nature is determined by the belief system it inherits. The two are not only interdependent; together they form a cognitive continuum, an extended and seamless knowledge, a world view, mostly nascent and occasionally expressed in songs and stories in oral traditions.

Michael Wessel's discussion of the /Xam sense of indignity (Chapter 6) brings home the organic fusion of nature and consciousness:

> The idea of the environment is a modern one, the product of a certain separation from life lived on the land and the formations of consciousness that attended the shift to urbanisation. Environmentalism is rooted in an

> awareness and politics of loss and conservation. /Xam thought, as it is reflected in the primarily pre-colonial world of the Bleek-Lloyd archive, did not possess a set of self-reflexive environmental principles.

The conflict triggered by the cultural, political and economic interventions implied in colonialism and the worldviews and value structures of indigenous peoples has resulted in a stereotyping of the indigenous as anthropology's "primitives." It has resulted in a severe loss of dignity for the indigenous – something they value more than any material possession. But most of all, these communities have had to pay the cost of modernity by losing their land, rivers and resources. Tracey Benson and Lee Joachin discuss in this volume (Chapter 4) how the creation of Canberra as the National Capital Region of Australia in 1967 required the blocking of a river. The essays included in this volume bring into discussion such instances form North and South America, Africa, New Zealand, Australia and India. Sadly for the indigenous, the post-colonial governments in most "free nations" have continued the colonial legacy of the clash between the indigenous and the "civilized" world. In her essay (Chapter 1), Archana Prasad analyzes how the "adivasis," a generic term used indigenously to designate the indigenous in India, have been a victim of the state repression. Prasad comments,

> The articulation of identities and the performance of rituals are conditioned by the material realities of adivasi life. Since the post-independence period, the adivasis have been subjected to different forms of dispossession through their adverse incorporation into capitalist modernity. This was largely structured through the state control of forests and the displacement of livelihoods.

My short piece, translated by Kavita Karnad (Chapter 2), is drawn from a book I published in Marathi a decade ago. The book was a result of my work with adivasis. During my work, I noticed how the rapid depletion of their forest resources had become a major cause of their pauperization. It is not as if there were not enough laws to protect their interests; but all of the law came to them as an alien system of thought and practice. The struggle to understand this alien sense of justice has taken the adivasis such a long time that by now they are almost at the fag end of their losing battle and left with a few broken English words as their only consolation. In that sense, the forest they lost now speaks English! I would like to believe that the perspectives brought together in this volume from all continents make it sufficiently useful in gaining more than an anthropological understanding of the situation of the indigenous.

In Chapter 3, Deborah McGregor discusses the impact of colonialism on water sufficiency in North America and points to how the indigenous knew in the past that the Great Lakes are an integral part of their existence and to manage the relationship as a unity between humans and nature. Charles Dawson et al. take up a variety of issues in Chapter 4 in the context of Australia and the Pacific, and establish that the knowledge of the indigenous did not begin with only herbs and hordes of animals, but extends to cover habitat, cosmology and governance understood in a much wider frame than the contemporary idea of politics and economy. In Chapter 8, Seth Garfield highlights the struggle of the indigenous in South America to contain the destructive impact of technology-driven development projects on the lives of the indigenous. Invariably, the indigenous are seen as a *non-grata* shadow from the past.

This five-volume series is intended to comment on the processes through which the clash of civilizations has played out and is impacting the society, culture, belief-systems and languages of the indigenous. Each of the volumes deals with a related set of two key issues crucial to understanding the indigenous and thinking about the processes affecting their culture and life. These are: *Environment and Belief Systems*, Volume 1; *Gender and Rights*, Volume 2; *Indigeneity and Nation*, Volume 3; *Orality and Language*, Volume 4; and *Performance and Knowledge*, Volume 5. These key concerns were selected for discussion based on my three-decade-long experience of living amidst and working with some of the indigenous communities in western India. A large number of consultations, discussions, workshops and field visits have led me to believe that at this juncture of history, these ten form the "key concepts" that one must understand in order to imagine and understand the indigenous peoples. The volumes present in-depth studies in the form of long essays, ranging from 8,000 to 10,000 words, and useful bibliographies. However, these essays are not exactly and purely academic studies. Their reference is not so much to the previously accumulated knowledge in the fields of anthropology, linguistics, literature, social sciences, law and art criticism; the essays focus on life lived more than on any field of knowledge. They relate to the prevailing contexts surrounding the communities discussed, without, however, lacking in academic rigour. Many of the contributors have been activists in addition to being scholars. Besides, they are drawn from all continents, in most cases from the communities themselves, and they bring to these volumes their valuable experience of the indigenous from all of those continents. Thus, the five volumes, focusing on ten key concepts, effectively speak about the indigenous peoples in Australia, New Zealand and the Pacific region; in India and East Asia; in Africa and in the Americas. Each of the volumes has seven or eight intensive essays which open up a range of themes and questions related to the specific key terms discussed. It is hoped that these volumes will form valuable reading for students, researchers and academics interested in knowing what the indigenous communities think about themselves and about the contemporary world.

The publication of this series of volumes brings me a great personal satisfaction. I was trained in literary studies in an era when "excellence" in literature was ascribed to works in the main European languages alone. In my early years as a professor at an Indian university, I started noticing that the rich and unique culture of the indigenous communities was at that time like a continent about to be submerged under the ferocious cultural assault of the urban-industrial values and materials. My unease increased so much that I felt compelled to drop out of academic life and to move to a tiny village where the Rathwa indigenous community lived. That opened a new universe for me. There was so much to experience, see and learn from them, most of all how limited what I had till then imagined as "knowledge" was. Throughout my years spent with them, it was never my intention to "represent" them to the rest of the world that was on a path of ecological destruction. My task was to let the indigenous express themselves, to create spaces for their voice, to facilitate that expression. I realized that the gap between the indigenous and the universe of formal knowledge, labour and economy was unbridgeable. During the last three decades, all of my intellectual and activist work has remained devoted to bridging this abysmal gap. The publication of this series, Key Concepts in Indigenous Studies, is a small step in that direction. I would like to hope in all humility that it will achieve what it aims to, and will remind the readers that the Earth does not belong to us, we belong to it.

Note

1 National Geographic, 'Resource Extraction and American Indians: The Invisible History of America', April 28, 2017, https://blog.nationalgeographic.org/2017/04/28/resource-extraction-and-american-indians-the-invisible-history-of-america/ (accessed 30 May 2019).

References

Platvoet, J.G. 1992. 'African Traditional Religions in the Religious History of Humankind', in G. ter Haar, A. Moyo and S.J. Nondo (eds.), *African Traditional Religions in Religious Education: A Resource Book With Special Reference to Zimbabwe* 11–28. Utrecht: Utrecht University.

World Bank, 'Indigenous Peoples', www.worldbank.org/en/topic/indigenouspeoples (accessed 29 May 2019).

1

RITUAL AND CULTURAL PRACTICE AMONG INDIAN ADIVASIS

Archana Prasad

Recent political developments and the large-scale onslaught of neo-liberal capital through its alliance with the forces of Hindu nationalism have forced scholars to relook at the problem of ritual and cultural practice in adivasi/tribal societies. The articulation of adivasi identity (which embodies within it a celebration of ritual and cultural practice) is not necessarily an alternative to these hegemonic and oppressive forces. Rather, as this essay shows, they may in fact get incorporated into dominant cultural practice. In order to explore this proposition, this essay reconstructs the contemporary history of the ways in which culture has been studied and practiced in adivasi life.

Apart from the introduction and conclusion, this essay is divided into six sections. The first two sections are conceptual, as they explore the theoretical perspectives and set up the argument to see ritual and cultural practice as socially and politically constructed entities. The third section further extends this argument by exploring the dialectical relationship between adivasi symbolism and their changing material reality. The last three sections locate the empirical history of adivasi politics and cultural practice within this context. Essentially, they look at the interaction between adivasi politics, cultural practice and the changing character of hegemonic forces. Within this structure, the essay argues for a more reflexive and complex way of understanding the relationship between class, culture and adivasi politics.

Religion, beliefs and "pre-capitalist" tribal societies

Since the advent of the Enlightenment, social scientists have viewed the belief systems of "indigenous communities" as "primitive" systems which are largely "uncivilized" in their lifestyles. This perspective espoused an idea of an ideal "civilized" modern society which was based on a modern rationality and where any form of unexplained belief was categorized as "superstitious." However, this evolutionary framework was contested by the development of sociology and anthropology – disciplines that tried to make sense of "tribal beliefs" within their social and economic contexts. Several anthropologists, like Evans-Pritchard or Malinowski, argued the tribal belief systems constituted a part of the functional organization of society (Evans-Pritchard, 1940). This means that religious beliefs and practices constituted a major part of the organization of the economy and society (Malinowski, 1948). The institutional framework in which this was done largely represented the customary authority which was embodied within kinship institutions. This approach, designated the "structuralist-functionalist" perspective, tried to explain the internal logic of these societies and argued that they had their own rationality. Seen in this context, tribal religion and beliefs are an essential part of a social organization that reproduces communitarian norms and institutions. Hence, the religion and belief systems were seen as a critical element of the survival of these communities themselves (Malinowski, 1948).

The juxtaposition between evolutionary and functional perspectives on tribal religion has formed the contours of the ideological debates. Liberal anthropological theories were often challenged by sociologists and anthropologists from the Marxist tradition (Seddon, 1978). The uniqueness of their analysis lay in distinguishing between "pre-capitalist" societies and capitalism. Simply put, theorists like Maurice Godelier and Meillassoux argued that the question of kinship and religion had to be seen in the context of the lack of the separation between the "base" and the "superstructure" within pre-capitalist societies (Godelier, 1977; Meillassoux, 1981). This implies that authority emanating from kinship and religion maintained the hierarchies and patterns of redistribution of wealth, eventually leading to processes of primitive accumulation. Further rituals and customs (such as marriage and gifts) were also methods of redistribution of wealth in tribal societies. Hence kinship, religion and custom became the ideologies of accumulation and extraction within pre-capitalist societies where the functions of production and social reproduction were conflated with each other because the there was no difference between the "base" and "superstructure" (Mauss, The Gift: Forms and Functions of Exchange in Archaic Societies, 1966). Thus, kinship and religion could be seen as ideologies of accumulation and control. This analysis is, however, not necessarily true for Marxist scholars within India, where scholars like D.D. Kosambi saw both the ideologies of kinship and caste as hegemonic ideologies which resulted in the adverse incorporation of tribal people into systems of extraction of labour and resources (Kosambi, 1965).

The location of culture and religion in relationships of power was implied in the analysis of Marxist anthropology, but was made more apparent within cultural theory. Culture was identified not in terms of rituals and symbolism, but as a way of life. Thus, Clifford Geertz wrote in the *The Interpretation of Cultures* that the study of cultures and religions is a two-stage process: "first, an analysis or the system of meanings embodied in the symbols which make up the religion proper, and, second, the relating of these systems to social-structural and psychological processes" (Geertz, 1973, 125). This definition has definite connotations when applied to adivasi societies. First, it implies that symbols are embedded in institutional structures with a certain mentality which informs the world view of adivasis and therefore also informs their politics. Second, this world view also creates cultural boundaries which structure the power structure within the community. Thus, as Bakhtin states, the performance of rituals and practice is in fact a method of the exercise of power. Alternatively, folk cultures represented the culture of protest where hierarchies were violated and social codes transgressed. Hence folk culture can be seen as a potential counter-hegemonic tool to enact opposition to existing communitarian structures (Bakhtin, 1984).

In the context of this discussion, it is now imperative to address the question as to whether the term "adivasi" has the same meaning as the term "tribal" in the Indian context. Several authors and researchers tend to use the term "tribal" and "Adivasi" interchangeably. It is also contended that the term adivasi is more authentic because it is a form of self-expression that expresses a community's own cultural and ethnic identity. The term "tribal" is considered more pejorative, as it implies a certain incorporation of the tribal elites within the existing political system (Rycroft and Dasgupta, 2011). These perspectives, however, treat non-class communitarian identities as the basic form of social organizations that exist within the political sphere and whose status under different stages of capitalism need not be a significant factor. Therefore, it is not surprising that all identities are treated by class-based movements, as both are dividing and hegemonic and supported by the modern capitalist state (Karat, 2011). But this perspective ignores the counter-hegemonic potential that many forms of self-expression may have acquired through a long historical process. This essay treats adivasi identity as an abstraction; a form of self-expression and representation whose material reality is formed by two concurrent processes. First are the processes of primary accumulation, where the alienation of social groups from their own resources integrates them within the lower end of the capitalist labour processes and strengthens the politics of difference (Prasad, 2010). Such a politics of difference is the result of a continuous process of class formation which determines the social basis and the ideological character of this identity. Therefore, it is not possible to attribute identities, in general, to any one form of class or elite politics. The meaning of "being adivasi" will be determined by the class character of collective or individual agency that seeks to organize these sections as agents of social transformation. In this sense, it is virtually impossible to have a static or an essentialist conception of "adivasi identity" whose form changes with the changing character of capitalist modernity (Prasad, 2004). Second, state-determined hegemonic tribal identities have emerged through the institutionalization of affirmative and protective measures which have become an important feature of bourgeois politics (Prasad, 2016a). Such hegemony has created its own opposition in the form of the politics of "adivasi identity" led by a stratum of educated tribal elite. The adivasi identity politics of this kind is oppositional and non-class in its orientation, has romantic neo-traditional influences and critiques all forms of modernity (Prasad, 2003, 2nd edition 2011; Damodaran, 2011). In this sense, both "tribal" and adivasi represent the generic forms of newly created political communities which re-create their own traditions at particular historical conjunctures. This invention of traditions is a legitimizing factor in the creation of new communitarian and national identities whose articulation will differ according to their social basis (Hobsbawm, 1990).

The question of belief systems has to be foregrounded in the previously mentioned observation because the way in which rituals and traditions are represented is itself intimately related with the material reality and the place of adivasi people in the larger political economy. The projection of folk adivasi culture as a counter to hegemonic capitalist modernity was enacted in a manner that drew cultural boundaries. These boundaries established the principle of hierarchy within the community and established certain customary norms which led to the exclusion of women and others from formal ritual practices within the community. They also reflected the morality and values of the adivasi society as it existed in their own times (Prasad, 2008b).

Customs, rituals and the construction of inequalities

As mentioned earlier, the construction of tradition and custom in adivasi societies is historically constructed and defines the culture of the communities vis-à-vis the dominant "other" (Bakhtin, 1984). However, in doing so, it itself acquires a hegemonic character which reproduces the inequalities within the communities. The construction of the rituals closely follows and represents the social structure of the adivasis. Broadly speaking, two types of social structures are identified in contemporary adivasi societies. The first are the patrilineal structures where inheritance and property rights are guided by patriarchal authority, and the exclusion of women and youth is evident in all kinds of ritualistic practices. Though these societies are largely non-idolatrous in character, the rituals performed during marriages and the invocation of gods during agricultural operations reflect the power hierarchy within the community. The second are the adivasi communities that follow matrilineal systems of inheritance. Though kinship and descent is traced through the mother, the husbands and brothers continue to enjoy significant powers. Hence, though these societies are matrilineal they are not matriarchal, where women enjoy real control and power.

Seen in this context, anthropological analysis of adivasi belief systems displays a certain type of romanticism in projecting a morally superior adivasi society. A good example of this is the work of Verrier Elwin who, in his magnum opus, *Baiga*, where he recorded the creation myths of the Baiga, also known as the high priest of the adivasi society in Madhya Pradesh and Chhattisgarh. As he records:

> The Baigas were established in the practice of bewar by Bhagvan himself, who when he called the tribes of the world to make a king first chose the Baiga. But Nanga Baiga begged that the Gond his brother be made king

> instead. Bhagvan was pleased, and as a mark of favour, took Nanga Baiga by the hand and placed him on his throne by his side. He granted his prayer to make the Gond King, but gave him a greater blessing.
>
> "All kingdoms of the world" he said "may fall to pieces but, he who is made of the earth and is Bhumiraja, lord of the earth shall never forsake it. You will make your living from the earth. You will dig roots and eat them. You will cut wood and carry it on your shoulders. Your wife will pick leaves and sell them. You must not tear the breast of your Mother the Earth with the plough like the Gond and the Hindu. You will cut down trees and burn them and sow the seeds in their ashes. But never become rich, for if you did you would forsake the earth and there will be no one to guard it and keep its nail in place."
>
> *(Prasad, 2008b, 317)*

This creation myth, as told by Elwin, establishes the authority of the Baiga patriarch as the protector and the creator of the world. The only mention of the woman, or the wife of Nanga Baiga, presents her to us as an ordinary gatherer of leaves, rather than one who is an equal partner in the procreation of the tribe or the world. What comes out supremely in the myth of the 1930s is the overpowering presence of the Baiga man as one who bears the burden of the entire world. However, if we trace the history of this myth from the time of the *Ain i Akbari,* perhaps this was not always so. A version of the creation story reproduced from the medieval text by ethnographers Russell and Hira Lal in *The Tribes and Castes of the Central Provinces of India* (1910) is the following:

> In the beginning they say, God created Nanga Baiga and Nanga Baigin, first of the human race, and asked them by what calling they would choose to live. They at once said they would make their living by living in the jungle, and permission being accorded, have done so ever since. They had two sons one of whom remained a Baiga, while the other became a Gond and a tiller of the soil. The sons married their own two sisters . . . and while the elder couple are the ancestors of the Baigas, from the younger are descended the Gonds and all the remainder of the human race.
>
> *(Russell and Hira Lal, 1910, 79)*

What is noticeable in the earlier version of the creation myth is the role accorded to both Nanga Baiga and his wife Nanga Baigin as the originators of the human race. Women, in this myth, are considered part of the ancestral heritage of the tribe. Perhaps the explanation for this lies in the way in which the history of bewar progressed in the last three centuries. The seasonal livelihood system of the tribe ensured that Baiga women had a better position in a livelihood system which was seasonal and multi-occupational in character. This was true because men and women contributed comparable amounts of labour time to the seasonal bewar cycle when it was connected to hunting and gathering practices, and the

Baiga had free mobility between different ecosystems (Prasad, 1998). This was reflected in the earlier myth which established the primacy of the Baiga male but did not altogether ignore the role of the woman.

However, new colonial forest laws led to the banning of bewar, and consequently, the Baiga were left with little option but to fight for its restoration. The successive events leading to the formation of the Baiga Chak in the late 19th century and resultant penury of its inhabitants placed an undue burden on women, as the main contribution of Baiga men to their livelihood was in the form of bewar (Prasad, 2003 2nd Edition 2011). At the same time, it also led to the aggressive articulation of the Baiga identity around bewar as seen in the myth recorded by Elwin. It is important to note that the main assets and meagre wealth of the Baiga emanated out of the land and implements used in bewar. And since the men controlled these assets, it is but natural that their role was exaggerated in an identity that was focused on bewar (Prasad, 1998).

It is important to understand that a bewar-centred identity was more patriarchal in character because of the structure of property rights and settlement patterns of the Baiga family. The identity of the family was also articulated contextually. The family consisted of the father and his sons sharing the same cluster compound. The patriarch was the head of the family and a *siyan* of the village. Under his supervision, all family disputes were resolved and marriages fixed. Relations between joint families were conducted through him. Once the father died the eldest son attained this position. The identity of the patriarch was only articulated in times of ritual performance and in the case of disputes with other families. In routine life, the family identity was constituted through the formation of work groups. The father, who was the head of the family, led his sons to clear the fields. The sons, who lived and dined with their father, did not have separate fields. According to community norms, when the son got married, he continued to work in the same field as his father and partake of the produce from the family land. As long as the son lived in the same compound, he belonged to the family work group. Before marriage even his axe belonged to his father, whereas after marriage he was gifted an axe by his father. In this situation the "axe," an essential element of bewar, was seen as a symbol of inheritance and it was considered impure if it was owned or used by "the woman." Thus, as in other patriarchal societies, the social position and right to inherit land, technology and property was through the male heir. It is therefore natural that any bewar-centred identity would enhance patriarchal authority (Prasad, 2008b). By the time Elwin recorded the myth the assertion of Baiga male authority was at its zenith, and the formal recording of the myth further accentuated this process. The Baiga is an example of the establishment of patriarchal authority by the ruling adivasi lineages. Though these lineages have lost their glory in the wake of the onslaught of capitalist modernity, the post-war era shows that men still control the narrative and its expression in terms of the larger belief system.

The example of the Baiga may, in some senses, be contrasted with the folklore of matrilineal communities. There are only three adivasi matrilineal communities:

Garo, Khasi and Jaintia. Their customary laws show that even though the status of the woman and the younger family that are not accorded to 'wives' in patrilineal societies. For instance, in Garo families, though the youngest daughter is the inheritor, she has no power to manage or alienate the land. The manager of the land is her husband or younger brother, who can also decide whether to sell off the land or not. Further, the norms of marriage are strictly observed where the heiress marries within the large clan in order to keep the property within the clan (Marak, 1986). Within the Khasi customary law, too, the youngest daughter is the custodian of property, but she is not seen as fit enough to manage her own property. Hence, the husband or the brother has special rights as manager of the property. Further, if the youngest daughter separates from her husband and marries outside the clan, she has to forego her inheritance rights (Cantile, 1934).

The constraints on the liberty of women are also reflected in the fact that almost no women are heads of community organizations within matrilineal societies. The head of the community organization is the man. Further, the Khasi woman also has no right in the religious sphere. An example is the funeral rituals within this matrilineal society which are largely carried out by the male member of the family. The heiress or the head of the family only plays a nominal supporting role. Similarly, in the case of the enactment of rituals to make a contract between "man" and "god," the high priest is the maternal uncle who is charged with the responsibility of the security of the family. Hence, it is quite clear that even though women were the custodians of the traditional family under the matrilineal system, they lacked the real power of control over their inheritance (Nongbri, 1984). The creation myth of the Khasis also gives precedence to the male as the original ancestor, and this special place is accorded special powers (Kharmawphlang, 2005). However, the exalted position of the heiress is quite evident in several ethical codes that have been evolved in matrilineal societies, thus displaying a different character of patriarchal authority.

As the previous discussion shows, in the case of both patrilineal and matrilineal societies, construction of identity through the enactment of rituals and the narration of tradition may have diverse manifestations, but it has a common morality which is developed through its interactions with other cultures. This morality sets up the distinction between purity and impurity, good and evil, and also structures the practices of conformity and non-conformity that go into the articulation of adivasi beliefs in the form of culturally particularistic identities (Douglas, 1966). But the striking similarities within these identities show that the belief systems have a shared material reality that informs their articulations.

Symbolic power and material reality

The articulation of identities and the performance of rituals are conditioned by the material realities of adivasi life. Since the post-independence period, the adivasis have been subjected to different forms of dispossession through their adverse incorporation into capitalist modernity. This was largely structured through the state control of forests and the displacement of livelihoods that made most adivasis dependent on wage labour (Prasad, 2010). It is thus not surprising that the adivasi leadership exalted its own history in order to mobilize people against oppression. In the process, they valorized the symbols and the rituals as an alternative to colonial modernity where the ideological hegemony of the Christian Missionaries, mainstream nationalists and the forces of right-wing Hindu nationalism structured the diversity of adivasi politics (Prasad, 2003 2nd Edition 2011). As the sociologists and historians of the time argued, most of these ideologies represented themselves in terms of Sanskritization models where the evolutionary progress of adivasis would take them into caste Hindu society. But, as D.D. Kosambi argued, this discourse was nothing but a way of getting access to the cheap labour of the adivasi people. It is therefore not surprising that one of the methods of the caste Hindu peasant economy was to incorporate the adivasi rituals into their own belief systems (Kosambi, 1965). In this sense, the performance of ritual was itself hegemonic in character since it integrates adivasis into oppressive social and economic relations through the development and transformation of power structures (Sinha, 1962) (Sinha, 1987). Later studies also emphasized some of these aspects through the valorization of the adivasi identity, which brought about some change of focus on the scholarship. Notable amongst these were the writings of the leaders of the movements themselves who emphasized their tradition (Munda and Bosu Mullick, 2003; Singh, 1983; Devalle, 1992)

The question of power has also been studied vis-à-vis the performance of rituals in terms of the structures of the rituals themselves. Though earlier anthropological works implicitly acknowledged the worth of symbolic capital in works like *Gift* by Marcel Mauss (Mauss, 1954), the role of symbolism as a source of power was more explicitly acknowledged by Pierre Bourdieu, who emphasized power as a strategic force in explaining and understanding the world (Bourdieu, 1990). The role of occults and priests within kin-based tribal societies must also be understood in the context of such power, where the power of knowing establishes customary authority. For example, even in matrilineal societies like the Khasis, the chief priest and his assistant are both males. An analysis of the rice myth and narrated ritual shows that women are still embedded in the traditional division of labour, where their basic role is to cultivate and prepare rice (Kharmawphlang, 2005). This role in caring for the field and the home is equated with caring for the "clan," Hence, despite rights inheritance, the ritual in fact displays the symbolic power of the chief priest, and in effect also reflects the domination of the husband and younger brother in decision-making within the Khasi household (as described before). The role of women in political institutions is also limited by such symbolic power (Goswami and Das, 2015).

Another example of the exercise of symbolic power is in storytelling and its recording. The choice of the informant and the anthropologist-recorder are crucial to projecting and legitimizing symbolic power. This is shown in the case of the Baiga, who are considered to be the priests of the Gond tribes in Central

India. Here, the role of the patriarch and the priest is legitimized through oral traditions which were captured in the famous text *Baiga* by Verrier Elwin. This fixing of the "myth" and "tradition" at a particular historical conjecture not only legitimizes patriarchal power, but also strengthens the social position of community elites. In this sense, the representation of the community in anthropological recordings plays an important part and a historical role in the formation of an aggressive patriarchal Baiga identity (Prasad, 2008b). Thus, the narratives of ritual practice themselves contribute to the perpetuation of symbolic power and its legitimization of the power relations within the material sphere. This relationship can be made by referring to the discussions in the previous sections.

Nationalisms and hegemonic religions in the 20th century

The adivasis of the 20th century have had to face ruling-class religions since the advent of colonialism. The main two competing religious orders can be identified as first, the Christian missionaries and second, the forces of Hindu nationalism. The rise of both these forces can be seen largely as a result of the inability of the mainstream nationalists to interpret the tribal question adequately. It is well known that the nationalists in Congress largely saw themselves as a modernizing force which sought to bring about the development of the "backward adivasis." In doing so, it privileged the culture of the mainstream Hindu society and advocated its reform so that the "adivasis" could be integrated within it in an appropriate way (Bose, 1941; Thakkar, 1941). This conception of reform was strongly influenced by Gandhi, who advocated constructive work amongst the adivasis. Within its overall framework, the mainstream nationalists sought to project a "unified" India so that the nationalist project of uniting different social groups could be achieved. It is therefore not surprising that the nationalists disassociated themselves from many "adivasi" movements, some of which interpreted nationalism in their own specific way (Baker, 1984). At a different level, the "adivasis" also had their own ways of understanding and integrating the nationalist sentiment into their belief system – a typical example of which was the Devi movement of Gujarat (Hardiman, 1987). This understanding embedded popular nationalist symbols within adivasi culture and interpreted them in specific ways that responded to regional and local forms of oppression.

The impact of the work of the church and the rise of Hindu nationalism amongst the adivasis has to be seen in this context. The Christian missions were amongst the earliest to set up institutions in both Central and Northeastern India. The American Baptist Church and the Scottish Free Church entered undivided Assam and Central Provinces, respectively. Though they started their work by opening up orphanages and providing famine relief, their expansion required them to become aware of the local customs and also recruit "native" adivasi priests. In the process, they impacted the rituals and narratives of the adivasi people. Though the missionaries were quick to denigrate the religion of the adivasis, in practice, many rituals were reinterpreted and given a new meaning in the local hues of Christianity. Thus, the unfolding of Christianity was itself was a complex phenomenon that entered the adivasi ritual practice in multiple ways. For example, the Khasi idea of the "supernatural" was transformed with the penetration of Christianity where all "supernatural" phenomena came to be explained as "miracles" (Lyngoh, 2015). A similar example can be seen from Jharkhand, where some of the rituals of the church were adopted by the Oraons, and their Karam festival was celebrated by the church (Kujur, 2010). Thus, there was a dialectical relationship between the church and adivasis where both religions adapted to each other, and where the church was not only forced to "indigenize itself," but also accept certain aspects of the belief systems (Robinson and Kujur, 2010).

One of the major impacts of this dialectical interaction was the sharpening of a distinct "Adivasi" or "tribal" nationalism in both Central India and the Northeast. This was particularly evident in the first decade of the 20th century. The interface between the church and the Mizo National Front (MNF) was strongly evident in Mizoram (Bhatia, 2010). Similarly, the agitation for a separate Naga state was attributed to the strong influence of the church (Nag, Expanding Imaginations: Theory and Practice of Naga Nation Making in Post-Colonial India, 2012). However, it would be fallacious to look at the role of the church in an instrumentalist fashion as a part of a larger imperialist design, as was seen by the mainstream nationalist movement. The political implications of the work of the missions can be attributed to a process of "modernization" and "awakening" that led to the formation of a new political identity. This was particularly due to the spread of missionary education that created a new social class within the adivasis which aspired both for social mobility and political power. Such a class created a powerful narrative of tradition and change which resulted in the contestations of the dominant conception of the "nation" (Nag, 2002).

This process was also evident in Central and Eastern India, where there was a strong demand for regional autonomy, the earliest being the demand for a separate state of Jharkhand by the Adibasi Mahasabha led by Jaipal Singh. The main feature of such articulations was the identification of a historical homeland which structured their demands. Just as the Nagas and Mizos demanded their control over their own customs and territories, the Adibasi Mahasabha, too, articulated the need to develop according to their own specific needs. As Jaipal Singh said in 1948, their "initiative has been of making the primitive man conscious of his political rights. Adibasis have begun to realize that their salvation is in their own hands" (Singh, 2000). This understanding laid the basis for the subsequent Jharkhand party and its demand for separate statehood that was achieved at the turn of the 20th century. This brand of nationalism lauded tribal rituals and beliefs as a part of a larger assertion of identity which was a major mobilizing force in political resistance.

Such a sub-nationalism, which was a result of the emergence of a new social class within the adivasi society due to the impact of different social forces, faced opposition from different forms of competing hegemonic forces. The first was the state itself through the process of the institutionalization of a "scheduled tribe" identity and through a process of the creation of designated "tribal areas" from the colonial period onwards (Prasad, 2003 2nd Edition 2011). This created a pathway for the incorporation of the emerging adivasi middle classes into the mainstream political culture. This resulted in the development of a ruling elite, which themselves preferred to fossilize cultures and beliefs and derive political capital out of them. In contrast, adivasi identity and culture developed in opposition to the official ruling-class version. It was more dynamic and politically volatile, thus getting a sharp response from the dominant caste Hindu society in the form of the rise of Hindu nationalism.

As is well known, the Hindu *Mahasabha* (the ancestral organization of the present day Hindu nationalists) was formed in the 1930s in Central India. One of its primary arguments was that the adivasis were "basically Hindus" and could not be considered as separate from the caste Hindu society. Hindu Mahasabha and its ideologues questioned the use of the word "adibasi" or "adivasi," and argued for the use of the term *banajati*. The main objection of leaders like K.M. Munshi lay in the fact that the "adibasi" did not exist as a single community, and the differentiation between communities had to be taken into account (Savyasaachi, 1998). But this understanding was not merely an analytical and empirical understanding of the formation of the adivasi identity, but was political in its character and was meant to prevent the consolidation of the politics of the "adivasi." The intellectual justification for this argument came from sociologists, like Ghurye, who claimed that the very concept of the "adivasi" was not based on historical truths and was imperialist in character because it was meant to divide Hindu society. Critiquing the census, Ghurye's main focus is on showing that the "tribes" were culturally similar and integrated within the diverse Hindu religion and its society (Ghurye, 1943). A classical case often cited in this regard is that of Siva, whose guards are adivasi because his originals are considered to belong to the adivasi society. This process was often termed as acculturation, but its analysts seldom discussed the terms of this process, which can be termed "adverse inclusion" into the "Hindu" society (Kosambi, 1965; Bose, 1941). Hence, we see that the birth of adivasi identity took place through a complex process, and its belief systems were posited in contrast with competing hegemonies, namely mainstream nationalism, Christianity and Hindu nationalism.

The celebration of adivasi culture and symbolism

The celebration and upsurge of an adivasi culture in contrast to the dominant cultures reflected a spirit of resistance, as well as the idea of an alternative to capitalist modernity. The full force of this symbolism came to the fore in the post-Nehruvian period, when the initial honeymoon between the tribal elites and the Indian state came to an end. The twin impacts of the emergence of a new, well-informed adivasi leadership because of affirmative action and the exploitative, uneven development arising out of the process of diversion led to the resurgence of adivasi identity politics. This politics became a part of a larger counter-hegemonic process that contested state capitalism on the question of rights and control over productive resources, on the one hand, and the assertion of cultural identity on the other (Heredia, 2002). This politics of adivasi rights focused on a radical critique of "modern development" and distinguished itself from the earlier class-based struggles over land and forest rights. In contrast to a class-based perspective, the adivasi identity movements advocated community control over local resources, as it identified the "modern" state as the main exploiter of the tribal people (Prasad, 2004). In this sense, the post-Nehruvian capitalist regime structured its own brand of primordial oppositional politics, setting the stage for a qualitative shift in the resource management regime which began to use the vocabulary of identity politics. Finally, the identification of the "Adivasi" with "indigeneity" was largely accomplished by the formation of the UN-sponsored Indigenous Peoples Working Group, in which "adivasi" elites were active (Roy Burman, 1994).

Within this broad context, the "adivasi" has been linked to the idea of culture and place. The first important symbol of adivasi politics was the image of the "adivasi" as the original inhabitant who had the historical right to their original homelands. This conception is part of the larger organic nationalism that considers control over resources as an essential part of self-determination and self-rule. Interestingly, it is not the political elites among the adivasis who have made this connection, but leaders like Guha Niyogi who linked the national question to the land question. Some scholars also term this struggle of the "original" inhabitants "ecological nationalism" (Cedelorf and Sivaramakrishnan, 2005). While rule over their own lands is crucial to the core conception of the "adivasi," it is important to underline that land not only refers to agriculture, but the links of agriculture with other resources, particularly forests.

Myths about the forests being the original abode of the "adivasis" have been revived from the writings of anthropologists and used to normalize the idea that the "adivasis" were the kings of the forests and had always lived inside them. In this sense, the idea of the "adivasi" reinforced the belief that they were people of the forests (Prasad, 2003 2nd Edition 2011). Significantly, such a process of marginalization in the pre-colonial period has been termed as "adjustment and coexistence" by pioneer environmental historians in India (Gadgil and Guha, 1992). The political motivation for such a construction, however, becomes clear when forest-dwelling "adivasis" rightly struggle for their legitimate rights over forest use at the height of state capitalism in forests after the 1970s. Hence, the myth of the "original inhabitant" is itself a creation of the process of primitive accumulation after the establishment of monopoly control over forests (Parulekar, 1975; Prasad, 2003, 2nd Edition 2011).

The second core value of "adivasi" ideology is the idea of a moral superiority (in terms of egalitarian and democratic values), and the cultural practices that form an essential part of identity. Of particular significance is the fact that the Jharkhand Mukti Morcha. under the leadership of Shibu Soren. mobilized ordinary "adivasis" to struggle by propagating the preservation of cultural values and practices that would act as markers of adivasi life. Hence, the political struggle for survival was seen as a cultural struggle to preserve the community (Munda and Bosu Mullick, 2003). In this sense, the creation of cultural boundaries that determine the limits of the political community is an essential component of being adivasi. Documentation of language and practices and an attempt to create an exclusive mode of communication that bound together several cultures were instruments used by the adivasi leadership for achieving this. However, such cultural symbolism almost always creates conflicts within the idea of the adivasi identity. Susana Devalle (1992) describes this process amongst the Hinduized adivasi communities and the Christian adivasis of Jharkhand. The sense of adivasi identity is much stronger amongst the Christian adivasis who have created their own literature and social spaces. In this regard, the reformist, as well as missionary, influences are well known. A good example of the protection of social spaces is the creation of *akharas*, or informal schools, by the Jharkhand Mukti Morcha in order to reproduce the practices and values that exemplified their identity. Similarly, the observance of Jharkhand Divas as a way of celebrating Jharkhand culture is again symbolic of the cultural protest that has come to represent the struggle for self-dignity and self-esteem (Devalle, 1992).

Within this framework, egalitarian social codes and customs are posited as alternatives to the unequal exploitative relations created by the modern state. Thus, harmony with nature and with humans is an essential characteristic of the ideology of such politics. Of particular interest is the projection of the status of women within this society in order to prove the moral superiority and greater democracy among the adivasis. The idea that women have freedom of mobility, the right to multiple marriages and co-equal powers within marriage has been suggested in several anthropological texts and has been reproduced within the adivasi image. Customary codes have been projected as being both prudent and egalitarian, even with respect to nature (Prasad, 2008b). This has helped to position the adivasi politics and image as an alternative to state-driven and controlled systems of conservation. Thus, the use of "adivasi" as a tool of mobilization and opposition to existing forms of domination has resulted in a cultural revivalism (Devalle, 1992).

The struggles for the creation of Jharkhand and Chhattisgarh in the last two decades of the 20th century were a reflection of this effort. They also reveal the hegemonic implications of the imagery that was used to create the social basis of the struggles (Prasad, 2003 2nd Edition 2011). It is well known that many of these struggles were led by the erstwhile ruling classes among the tribal people, and this was reflected in the nature of the symbolism itself. For example, the dominance of Mundari symbolism, Santhali vernacular vocabulary and several other social customs of larger tribes tends to obfuscate the voices of smaller tribal groups (Devalle, 1992). Further, the apparently egalitarian character of the adivasis does not take the trials, tribulations and burdens of adivasi women into consideration (Prasad, 2016b). Any acknowledgement of this would only expose the fault lines of the imagined adivasi community that has achieved its limited objective of obtaining a separate Jharkhand and Chhattisgarh state at the turn of the century. In this sense the celebration and revival of adivasi culture can be considered hegemonic and prone to making critical compromises with processes leading to adverse inclusion of adivasi ruling classes into the dominant political structure (Prasad, 2016a). This process also structures the recent intensification of the incorporation of adivasis into dominant social and economic structures.

Patterns of adverse integration into "mainstream Hinduism"

For 60 years, it was assumed that the belief systems of the adivasis were part of their "backward" social structures that needed to be reformed. However, the revival of the Jan Sangh and the activities of the Rashtriya Swayamsevak Sangh (RSS) in the post-emergency phase started a new pattern of domination. In order to expand itself, the Bharatiya Janata Party (BJP), a new party affiliated to the RSS, was borne with the aim of building a "unified and harmonious Hindu society." This process of unification has been essential in laying the foundations of a "Hindu state" which would represent all sections of society and bring about a polarization that would contest the claims of a "modernist" vision of a nation. To this end, the RSS revived its Vanvasi Kalyan Ashram and started the project of "reconversions" of Christian adivasis. Though this drive and the targeting of Christian adivasis can be traced back to the early independence era, it took new forms from the late 20th century onwards. Two trends could be noticed in the manner in which Hindu nationalists attempted to incorporate the adivasis into their fold: the first was reconversion; and the second, the penetration of adivasi identity, which itself became a force of polarization. It is therefore obvious that the forces of Hindu nationalism themselves realized the limits of their reconversion programmes and started to adapt themselves to regional contexts.

The first regional trend that was starkly evident was in the form of the revival of an aggressive "reconversion campaign" in Western, Eastern and Central India. The 1980s saw a dual trend in these three regions, where the institutional spread of the Vanvasi Kalyan Ashrams has structured the inclusion process. The Ashram was formed in Jashpur by Ramakant Deshpande and the RSS and Motubhau Ketkar with the help of the Maharaja of Jashpur. The Ashram was explicitly to bring adivasis back to "Indian culture" (Pachpore, 2013). In order to fund itself and expand its outreach, the Ashram actively collaborated with sadhus, yogis and traders. The blessing of the sadhus gave it legitimacy to fund the activities of the Ashram. The traders saw the Ashram as a counter to the rise of adivasi identity politics that often protested against the exploitation of adivasis by the market (Sundar, 2009). The Ashram used the vehicle of *ekal vidyalayas* and *shishu mandirs* to promote the culture of caste Hindu societies. The process by which adivasi children are socialized and disciplined into a "moral code" of Hindu society has been explained by analysis of these schools in Central India (Froerer, Religious Division and Social Conflict: The Emergence of Hindu Nationalism in Rural India, 2007b) (Froerer, 2007a). But quite apart from this, there has also been another important dimension of the educational work done by the Ashram. Residential schools were set up and tribal children from Northeastern states were educated in the "Hindu way of life" in these hostels. The main aim of these hostels is to do "dharma jagran" and make the adivasis realize that they are part of Hindu society (Sundar, 2002). One of the important components of such an education is to promote purist notions about "womanhood" and "girlhood," and therefore project the need to preserve the cultural unity of the "Hindu" society (Manjrekar, 2011). Gujarat also saw a large spate of anti-Christian attacks from the late 1990s which formed the basis of the polarization which led to the participation of some adivasis in the Gujarat riots of 2002 (Puniyani, 2006). The socialization of adivasis as "Hindus" was also seen in the case of Odisha, where a systematic campaign was carried out between the Christian adivasis and others in order to create conditions of conflict. This conflict ultimately led to the Kandhmal riots of 2008, and demonstrated the role of the Vanvasi Kalyan Ashram's in creating such polarization (Prasad, 2008a; Chatterjee, 2009). In all these regions, the "Hindu way of life" was ritualized and the tribal culture was incorporated as a peripheral culture (Froerer, 2007b)

This march of Hindutva in Central, Eastern and Western India was accompanied by the march of capital. The Hindutva forces were often seen as aligning with capital and making way for them. Of particular interest in this regard is the spearheading of the counter-insurgency anti-Maoist operations in Chhattisgarh. The advent of these operations under the banner of Salwa Judum also saw an increase in the anti-Christian attacks within the state. This was accompanied by the opening up of natural resources to mining companies and the coordination between the operations of security forces and the companies (Prasad, 25 June 2016). These emerging conflicts have only paved the way for the domination of Hindutva forces over adivasis.

The second important trend in the expansion of Hindutva has been the radicalization of adivasi identity itself, especially of the dominant lineages. This has been particularly evident in the Northeastern region, where the RSS has been trying to spread its wings. The sharpening of the narrative against "illegal Muslims" and the intervention of Hindutva forces in Bodoland created a situation of social conflict. As a BJP MP remarked during the Bodo action against the Muslims in 2012, "this [the Kokrajhar riots] is not a communal conflict, but a conflict between Indians and foreigners." In a similar vein L.K. Advani stated in the Lok Sabha that, "this is not a Hindu-Muslim issue, even though there may be some truth in the matter. The main issue is who is an Indian and who is a foreigner. The government must decide this and also deport the Bangladeshis." Thus, Tarun Vijay, a part of the BJP's delegation to Kokrajhar, writes in *The Organiser* (August 13, 2012) that this "is time for Hindu society to ponder over their decline and why the foreigners have gained so much of power to attack them in their own land." This appropriation of the Bodos into the Hindu fold is not surprising and is being used to legitimize the intervention of fundamentalist groups in order to expand their social and political basis amongst tribal elites. The vulnerability of the Bodoland politics to such appropriation is evident in the public utterances of the leaders of the Bodoland Territories Autonomous Districts (BTAD). They have often been quoted as saying that they have been "invaded by the Bangladeshis" and routed from their own homes. They have further asserted that they would not allow the Muslim refugees to return to their homes unless they have verified their "citizenship." Such an assertion has only fed into the myth of the foreign invasion that has been repeatedly used by the RSS to expand its influence (Prasad, 2012). A news report noted that the use of identity politics to spread "nationalist values" in the Northeast was being done through the spread of ekal vidyalayas and Vanvasi Kalyan Ashrams, whose number has risen to more than 5,000 in Assam alone. The practice followed in these schools is one that promotes North Indian "Hindu Sanskriti," even though their ideologues seldom contradict the tribal culture as practiced by tribal elites (Siddiqi, 2014). In this sense, the politics of tribal/adivasi identity is not always in contradiction to the spread of "Hindu culture and morality," and should not be necessarily seen as a form of radical politics.

By way of a conclusion

The preceding discussion has highlighted the changing nature of cultural practices and ideologies in adivasi/tribal life in different regions of India. Conceptually, it has seen the question of ritual practice and culture as embedded within the larger socio-political and material context. In this sense, the sphere of culture and ritual practice is neither autonomous nor necessarily a realm that is independent of the larger development of capitalism. Contemporary perspectives on ritual practices and culture need to be seen in this context. This essay has shown that rituals and cultural practice are socially and politically constructed. In doing so, it has addressed the nature of changing identities and the articulation of cultural types through tropes of resistance and political expression.

Within this overall conceptual framework, the essay points towards the emerging challenges for tribal/adivasi culture in all regions of the country. The contemporary hegemonic onslaught by the alliance of Hindu nationalism and neo-liberal capitalism is forcing the supporters of communitarian politics to rethink their strategies. This is largely because the spread of Hindu nationalism has itself changed the character of adivasi/tribal societies, and the clash between the ruling classes and the labouring poor has increased (Prasad, January 2014). Hence, the social basis of the politics of the adivasi will determine whether it succumbs to Hindutva politics or resists the forces aligning with the greatest exploiters of the adivasi people (Prasad, 2016a). Thus, the real challenge is to unpack the relationship between class, culture and the politics of the adivasi in contemporary India.

Bibliography

Abrol, D. (2006). *National Environmental Policy, 2006: Liberalising Environmental Regulation*. New Delhi: Dinesh.

Andhra Pradesh Revolutionary Communist Committee. (1969). *Immediate Programme*, April 10–12.

Anon. (2005). "Forests and Tribals" (Special Issue). *Seminar*, August.

Baker, D. (1984). "A Serious Time: Forest Satyagraha in Madhya Pradesh, 1930". *Indian Economic and Social History Review*, 21(1): 1–11.

Bakhtin, M. (1984). *Rabelais and His World*. Bloomington: Indiana University Press.

Banerjee, S. (1984). *India's Simmering Revolution: The Naxalite Uprising in India*. London: Zed Books.

Bhatia, L. (2010). "Contradiction and Change in the Mizo Church". In R. Robinson and J. M. Kujur (eds.), *Margins of Faith: Dalit and Tribal Christianity in India*. New Delhi: Sage Publications.

Bose, N. K. (1941). "The Hindu Method of Tribal Absorption". *Science and Culture*, 7: 188–194.

Bourdieu, P. (1990). *The Logic of Practice*. Stanford, CA: Stanford University Press.

Cantile, K. (1934). *Notes on Khasi Law*. Khongpai Publisher.

Cedelorf, G., and Sivaramakrishnan, K. (2005). *Ecological Nationalisms: Nature, Livelihoods and Identities in South Asia*. New Delhi: Permanent Black.

Chatterjee, A. (2009). *Violent Gods: Hindu Nationalism in India's Present; Narratives from Orissa*. New Delhi: Three Essays Collective.

Communist Party of India (Maoist). (2007). *Strategy and Tactics, Document Adopted by Unity Congress*.

Communist Party of India (Marxist). (1967). *On Left Deviation: Resolution Adopted by the Central Committee*. New Delhi: CPI (M), August.

Communist Party of India (ML-People's War). (2000). *New People's Power in Dandakaranya: A Report on the Ongoing Developmental Activities Undertaken through the New Organs of Power, Led by the CPI(ML)* [People's War].

Communist Party of India (ML-People's War). (2001). *Krantikari Jan Kamitiyon Ke Bare Mein*.

Comptroller and Auditor General of India. (2013). *Compliance Audit on Compensatory Afforestation in India, Union Government, Ministry of Environment and Forests*. New Delhi: Government of India.

Damodaran, V. (2011). "Customary Rights and Resistance in the Forests of Singhbhum". In D. Rycroft and S. Dasgupta (eds.), *Politics of Belonging*. London: Routledge, 103–118.

Devalle, S. (1992). *Discourses of Ethnicity: Culture and Protest in Jharkhand*. New Delhi: Sage Publications.

Douglas, M. (1966). *Purity and Danger: An Analysis of Concepts of Pollution and Taboo*. London: Routledge and Kegan Paul.

Ekta Parishad. (1999). *Report of the Joint Mission on Madhya Pradesh Forestry Project*. Bhopal: Ekta Paishad.

Evans-Pritchard, E. (1940). *The Nuer: A Description of Modes of Livelihood and Political Institutions of Nilotic People*. Oxford: Clarendon Press.

Froerer, P. (2007a). "Disciplining the Saffron Way: Moral Education and the Hindu Rashtra". *Modern Asian Studies*, 41(5): 1033–1071.

Froerer, P. (2007b). *Religious Division and Social Conflict: The Emergence of Hindu Nationalism in Rural India*. New Delhi: Social Science Press.

Gadgil, M., and Guha, R. (1992). *This Fissured Land: An Ecological History of India*. New Delhi: Oxford University Press.

Geertz, C. (1973). *The Interpretation of Cultures: Selected Essays by Clifford Geertz*. New York: Basic Books.
Ghurye, G. (1943). *Aborigines: So Called and Their Future*. Pune: Gokhale Institute of Politics and Economics.
Godelier, M. (1977). *Perspectives in Marxist Anthropology*. Cambridge: Cambridge University Press.
Goswami, N., and Das, S. (2015). *Community Rights and Women in Khasi Society, In Women Empowerment in Northeast India*. Ed. Sangeeta B. Tamuli. Guwahati: Eastern Book House.
Hardiman, D. (1987). *The Coming of the Devi: Adivasi Assertion in Western India*. Delhi: Oxford University Press.
Heredia, R. (2002). "Interrogating Integration". *Economic and Political Weekly*, 37(52): 74–78.
Hobsbawm, E. (1990). *Nations and Nationalism since 1780*. London: Cambridge University Press.
Karat, P. (2011). "The Challenges of Identity Politics". *The Marxist*, 27(1–2).
Kharmawphlang, D. (2005). "The Ka Lukhmi: A Khasi Rice Myth". *India International Centre Quarterly*, 32(3/2).
Kosambi, D. (1965). *Culture and Civilisation in Ancient India*. London: Routledge and Kegan Paul.
Kujur, J. M. (2010). "Tribal Church in the Margins: Oraons of Central India". In R. Robinson and J. M. Kujur (eds.), *Margins of Faith: Dalit and Tribal Christianity in India*. New Delhi: Sage Publications, 29–50.
Lyngoh, M. (2015). *Transformation, Tradition and Lived Realities: Vernacular Belief Worlds of the Khasis of Northeastern India*. University of Tartu Press.
Malhotra, K. (1997). "Joint Management of Forest Lands in West Bengal: A Case of Jamboni District". In S. B. Roy (ed.), *Policy to Commissioning of Joint Forest Management*. New Delhi: Inter-India Publications.
Malinowski, B. (1948). *Magic, Science and Religion and Other Essays*. Boston: Beacon Press.
Manjrekar, N. (2011). "Ideals of Hindu Girlhood: Reading Vidya Bharati's *Balika Shikshan*". *Childhood*, 18(3): 350–366.
Marak, J. (1986). *Garo Customary Laws and Practices: A Sociological Study*. Calcutta: Firma KLM.
Mauss, M. (1954). *The Gift: Form and Reason for Exchange in Archaic Societies*. London and New York: Cohen and West.
Mauss, M. (1966). *The Gift: Forms and Functions of Exchange in Archaic Societies*. London: Cohen and West.
Meenakshi Mukherjee, K. K. (2009). *Environmentally Sensitive Areas*. Pune: Kalpavriksh.
Meillassoux, C. (1981). *Maidens, Meals and Money: Capitalism and the Domestic Community*. Cambridge: Cambridge University Press.
Ministry of Environment and Forests. (2006). *National Environment Policy*. New Delhi: Ministry of Environment and Forests, Government of India.
Ministry of Tribal Affairs. (2013). *Progress Report of Implementation of Forest Rights Act as of 30.9.2013*. New Delhi: Government of India.
Munda, R., and Bosu Mullick, S. (2003). *The Jharkhand Movement: Indigenous People's Struggle for Autonomy in India*. Copenhagen: IWGIA.
Nag, S. (2002). *Contesting Marginality: Ethnicity, Insurgency and Sub-Nationalism in North-East India*. New Delhi: Manohar Publishers.
Nag, S. (2012). "Expanding Imaginations: Theory and Practice of Naga Nation Making in Post-Colonial India". *South Asian History and Culture*, 33(2): 177–196.

National Sample Survey Organisation. (2001). *Report of Employment and Unemployment Situation by Social Group, 1999–2000, Report 469*. New Delhi: Ministry of Statistics and Programme Implementation, Government of India.

National Sample Survey Organisation. (2006). *Report of Employment and Unemployment Situation by Social Groups, 2004–05, Report 516*. New Delhi: Ministry of Statistics and Programme Implementation, Government of India.

National Sample Survey Organisation. (2012). *Report of Employment and Unemployment Survey by Social Groups, 2009–10, Report 543*. New Delhi: Ministry of Statistics and Programme Implementation, Government of India.

Nongbri, T. (1984). "Gender and Khasi Family Structure: Some Implications of the Meghalaya Succession to Self-Acquired Property Act". *Sociological Bulletin*, 37(1/2): 71–82.

Pachpore, V. (2013). The Legend Called Balasaheb Deshpande: 'The Legend Called Balasaheb Deshpande' by Virag Pachpore, December 23, at http://organiser.org/Encyc/2013/12/23/The-legend-called-Balasaheb-Deshpande.aspx?NB=&lang=4&m1=&m2=&p1=&p2=&p3=&p4=&PageType=N

Parulekar, G. (1975). *Adivasis Revolt: The Story of the Warli Peasants in Struggle*. Kolkatta: National Book Agency.

Poffenberger, M., and McGean, B., eds. (1996). *Village Voices, Forest Choices: Joint Forest Management in India*. New Delhi: Oxford University Press.

Prasad, A. (1998). "The Baiga: Local Economy and Survival Strategies in Colonial Central Provinces". *Studies in History*, 14(2), December: 325–348.

Prasad, A. (2003 2nd Edition 2011). *Against Ecological Romanticism: Verrier Elwin and the Making of an Anti-Modern Tribal Identity*. New Delhi: Three Essays Collective.

Prasad, A. (2004). *Environmentalism and the Left: Contemporary Debates and Future Agendas*. New Delhi: Leftword Books.

Prasad, A. (2005). *Tribal Livelihood and Globalisation: Exploring the Potential of Non Timber Forest Produce in Central India*. New Delhi: Nehru Memorial Museum and Library.

Prasad, A. (2006). "Conservation and Tribal Development". *Social Scientist*, 34(7), July–August.

Prasad, A. (2006). "Re-Engineering Environmenal Impact Assessment". *People's Democracy*, September 24.

Prasad, A. (2006–2007). "Survival at Stake". *Frontline*, 23(26), December 30–January 12. New Delhi: The Hindu.

Prasad, A. (2007). "Marginal People and the Politics of Anti-Modern Development". In R. Chakrabarti (eds.), *Situating Environmental History in India*. New Delhi: Manohar.

Prasad, A. (2008a). "Kandhmal: The March of Hindutva in Tribal Orissa". *People's Democracy*, October 12.

Prasad, A. (2008b). "Patriarchies of the Community: Women and Oral Traditions in Verrier Elwin's Baiga". *Indian Historical Review*, 35(2).

Prasad, A. (2009). "On the Margins of Indian Planning". In V. Upadhyay, et al. (eds.), *Two Decades of Neoliberalism: The Development Process in India*. New Delhi: Danish Books.

Prasad, A. (2010). "Capitalism, Forestry and Tribal Labour in Central India". *Social Action*, 60(2): 138–151.

Prasad, A. (2010a). "The Political Economy of 'Maoist Violence' in Chhattisgarh". *Social Scientist*, 38(3–4): 3–24.

Prasad, A. (2010b). "Neo-Liberalism, Tribal Survival and Agrarian Distress: The Experience of Two Decades of Economic Reforms". In Alternate Economic Group (ed.), *Two Decades of Neo-Liberalism: The Alternative Economic Survey*. New Delhi: Daanish Books.

Prasad, A. (2012). Foreigners, Hindutva and Bodoland Politics. *People's Democracy*, August 26.
Prasad, A. (2014). *Class, Community and Identity: Politics of the 'Adivasi' in Contemporary India*. New Delhi: Oxford University Press.
Prasad, A. (2014). "Structural Changes in Tribal Societies: Evidence from 'Four Least Developed States'". *Yojana*, 58, January. New Delhi: Planning Commission, Government of India.
Prasad, A. (2016). "Adivasis and the Anatomy of a Conflict Zone". *Economic and Political Weekly*, 51(26), June 25: 27.
Prasad, A. (2016a). "Adivasis and the Trajectories of Political Mobilisation in Contemporary India". In M. Radhakrishna (ed.), *First Citizens: Studies on Adivasis, Tribals and Indigenous People in India*. New Delhi: Oxford University Press.
Prasad, A. (2016b). "Adivasi Women, Agrarian Change and the Forms of Labour in Neo-Liberal India". *Agrarian South: Journal of Political Economy*, 5(1).
Puniyani, R. (2006). *The Politics of Anti-Christian Violence: A Compilation of Investigation Committee Reports on Violence against the Christian Minorities*. New Delhi: Media House.
Robinson, R., and Kujur, J. M. (2010). *Margins of Faith: Dalit and Tribal Christianity in India*. New Delhi: Sage Publications.
Roy Burman, B. K. (1994). *Indigenous and Tribal People: Gathering Mist and New Horizon*. New Delhi: Mittal Publicatons.
Russell, R., and Hira Lal, R. B. (1910). *Tribes and Castes of the Central Provinces of India* (Vol. 2). Nagpur: Government Press.
Rycroft, D., and Dasgupta, S. (2011). *The Politics of Belonging: Being Adivasi*. London: Routledge.
Samata and CRY Net. (2001). *Joint Forest Management: A Critique Based on People's Perceptions*. Vishakhapatnam: Samata and CRY Net.
Sanyal, K. (1978). "Report on the Peasant Movement in the Terai Region". In D. P. Samar Sen (ed.), *Naxalbari and after: A Frontier Anthology*. Calcutta: Kathashilpa.
Sarin, M. (2001). "Defenders of the Forests: Disempowerment in the Name of 'Participatory Forestry'". *Forest, Trees and People*, (44), April.
Sarin, M., Singh, N. M., Sundar, N., and Bhopal, R. K. (2003). "Devolution as a Threat to Democratic Decision-Making in Forestry. Findings from Three States in India". In D. Edmunds and E. Wollenberg (eds.), *Local Forest Management: The Impacts of Devolution Policies*. London: CIFOR and Earthscan, 55–126.
Savyasaachi. (1998). *Tribal Forest Dwellers and Self Rule: The Constituent Assembly Debates in India*. New Delhi: Indian Social Institute.
Schreckenberg, K., Marshall, E., and te Velde, D. W. (2006). "NTFP Commercialisation and the Rural Poor: More Than a Safety Net?". In E. Marshall et al. (eds.), *Commercialisation of Non-Timber Forest Products*. Cambridge: UNEP and WCMC, 71–76.
Seddon, D. (1978). *Relations of Production: Marxist Approaches to Ecoomic Anthropology*. London: Frank Cass.
Shah, A. (2013). "The Agrarian Question in a Maoist Guerrilla Zone: Land, Labour and Capital in the Forests and Hills of Jharkhand, India". *Journal of Agrarian Change*, 13(3): 424–450.
Shankar, P. (2006). *Yeh Desh Hamara Hai: Dandkaranya Ke Karantikari Andolan Ka Itihas*. New Delhi: New Vista.
Shankar, R. B. (2000). *Van Dhan Scheme in Central Bastar*. Jakarta: CIFOR.
Siddiqi, F. A. (2014). "Target North East: How the RSS Plans to Make the Region Saffron". *Hindustan Times*, December 15.
Singh, J. (2000). "'Jai Jharkhand, Jai Adibasi, Jai Hind', Address to the All India Adivasi Sabha in 1948". In R. D. Munda and S. B. Mullic (eds.), *The Jharkhand Movement: Indigenous People's Struggle for Autonomy in India*. Copenhagen: IWGIA.

Singh, K. (1983). *Birsa Munda and His Movement 1872–1901: A Study of a Millenarian Movement in Chhotanagpur.* New Delhi: Oxford University Press.
Singh, P. (1995). *The Naxalite Movement in India.* New Delhi: Rupa and Co.
Sinha, S. (1962). "State Formation and Rajput Myth in Central India". *Man in India,* 1(42): 20–35.
Sinha, S. (1987). *Tribal Polities and State Formation in Pre-Colonial Eastern and North East India.* Calcutta: K.P Bagchi and Company.
Sudhakar, A. (2006). "A Sag of Twenty Five Years of Glorious Struggle: An Epic of People's Transformation". *People's March,* 7(1): 3.
Sundar, N. (2002). "Indigenise, Nationalise, Spiritualise: An Agenda for Education?". *International Journal of Social Science,* 173: 373–383.
Sundar, N. (2009). "Adivasis versus Vanvasis: The Politics of Religious Conversions in India". In D. Kumar and Y. Sunny (eds.), *Proselytisation in India: The Process of Hinduisation in Tribal Societies.* New Delhi: Aakar Books.
Thakkar, A. (1941). *The Problem of the Aborigines in India.* Pune: Kale Memorial Lecture, Gokhale Institute of Politics and Economics.
Vani, M. (2009). *Nought Without a Cause: Almost Everyone's Guide to the Underlying Causes of Deforestation and Forest Degradation in the Era of Neo-Liberal Globalisation.* Pune: Kalpavriksh and Global Forest Coalition.
World Bank. (1991). *Forest Strategy and Forests: Poor Countries.* Washington, DC: World Bank.

2

FORESTS NOW SPEAK ENGLISH

The indigenous at odds with the state

G. N. Devy[1]

My maternal grandfather resided in a locality named "Timber area" in Sangli, a city in South Maharashtra. He used to assist architects and builders employed by the feudal prince Patwardhan who ruled the small state from Sangli. The elegant colonial houses built by them had a generous use of wood. The city had several large warehouses of timber. The supply of wood came from trees felled in the forests surrounding Alnavar in Karnataka. These trees had large trunks and were ramrod straight. At my tender age, I could guess with great accuracy how recently had a particular tree been felled from their aroma. My grandfather died when he was 90. His sons, my maternal uncles, continued the same profession for another three to four decades after he stopped looking after it. Later, however, these trees were slowly replaced in those warehouses by glass sheets, steel and plywood. In their seven or eight decades of business, the lush forests of Alnavar were probably already decimated.

My grandfather owned a large wooden shelf full of books. Among the books, there was a magazine called *Chandoba*, sporting colourful pages. In these pages, I first met a heroic character named "Tarzan" who was fictionally located in the dense forests in Africa. I read all the stories of Tarzan which were published in those magazines right from the beginning to the end. This mythical forest dweller, a great friend of wild animals, an enthusiast, left an indelible impression on my tender mind.

My paternal grandfather was engaged in gathering forest nuts from the low valleys and hills in close vicinity of his village Bhor. The "permit" to collect these herbs was granted to him by the British District Collector, whose office was located at the district headquarters in Pune. After grandfather expired, the "permit" remained where it was kept among other documents. All of these documents were pierced with a sharp-edged wire, and the wire hung by a loft in the house. This was his filing system. The forests from where he collected wild nuts became real through those documents for my mind as a child.

During Grandfather's time, a dam was constructed – the Bhatgar Dam – on the river Nira, which flowed by the village. The purpose was to set up a powerhouse for the generation of electricity. A factory named R.L. Mills was started near the dam within a few years. The factory produced dyes from the nuts my grandfather used to collect. Small ferry boats would carry these forest products from the woods to the factory.

My father was a close follower of a saint in another district in Maharashtra. This association led to Father meeting another devotee – a man from Pune who was the owner of a paper mill. In 1980, I had just started teaching at the Surat University in Gujarat. About that time, there was an occasion to visit this paper mill. It was located at Vyara near Songarh, known at one time for its dense forests. The paper produced in this mill was used for several decades for printing books that contained literature in Marathi and Gujarati, the two major languages in western India.

As a young reader, at one time I was charmed by the works of Jim Corbett, who wrote about tigers and forest life in the lower Himalayas. Later, when I visited Nainital, I realized that none of the forests he described existed as they should, though one of them is named Corbett Park.

In my grandfather's massive bookshelf in Sangli, I chanced upon a Marathi translation of a biography of Abraham Lincoln. It mentioned that Lincoln lived in a town named Springfield during his early years. I felt that his being a timber-breaker during his youth was an irony of sorts. During my college days, the Western cowboy movies were very popular. I was enamoured by the treeless landscape of the Midwest. I would always wonder if America was as "shadeless" even when Columbus first stepped on its shore.

After completing my Ph.D. at the Shivaji University at Kolhapur, I went to Leeds University at England for some more study. There is a massive library named the Brotherton library on the campus of the university. The wood panels on all of its walls were very attractive. But the trees in the forests around England appeared quite old. I used to wonder, if the local trees were so well protected, where was the timber for the Brotherton library procured? I visited Leeds several times, but I had to wait to get an answer till I visited America some years later.

On the East Coast, a university named Yale is situated in a small town called New Haven. I had gone to Yale for research work. There, I became aware that Emmanuel Yale was the person after whom the university is named. Born in England, he lived most of his life in India and had acquired the contract of felling huge tracts of forest lands. He was known as a parsimonious and impulsive person. The enormous amounts of money earned from this business were later donated by him to the college in New Haven. In acknowledgement of his generosity, the college was named after him. Two centuries later, the university continues to give the United States most of its presidents.

While writing my first book in literary criticism, *After Amnesia*, I chanced upon a book on the history of the railway in Canada. This study was rich in statistics and had outlined how the laying of the railway tracks had devastated the indigenous forests. This made me extremely restless.

At Baroda, where I taught literature, my engagement with comparative literature took me to the study of French literature. During this phase, I noticed that the protagonist of Victor Hugo's classic *Les Misérables*, who comes up from the depths of poverty and manages to escape from a prison for criminals, later becomes the mayor of a small town. All the stages of this hero's life were intimately connected with wooden objects like carts, boats, etc. Rousseau, the French thinker, appealed to humanity to "return to nature." I thought, one may not fully understand why Rousseau argued so until one reads Victor Hugo.

Indian poet and translator A.K. Ramanujan has a short story in which a man escapes the orthodox Hindu society in Mysore and runs away to America for a higher education. Having put his past behind him, he comes to know of his father's death many years later – only when he chances upon a photo of his widowed mother in a book. He sees a photograph with her head shaved after widowhood in an anthropology book; he discovered in a distant land the entire tradition he already knew from his childhood, but did not know that he knew it. I should have used my childhood experience of the dwindling forests near the Bhor village and Alnavar woods for constructing the history of the colonial times. However, this realization came to me through the history of Yale University and the wooden panels in the Leeds University Library.

Since the time of Columbus, there was so much deforestation in America that during the time of Yale (in the latter half of the 18th century), wood had to be imported from other countries. Where did all those indigenous American trees go? For the answer to this question, one has but to look at the all the log houses in every American town. In the colonial era, England, France and Germany laid the entire railway network with wooden sleepers. These countries also invested in creating a massive navy and imposing ships. In most European countries, houses have wooden flooring so as to maintain the internal temperature. However, this practice led to the destruction of most of the forests in the colonies. After print technology created the printing industry, and after accepting that written documents became the evidence of truth, the world began consuming more paper. This further led to more and more trees being chopped. The colonial masters built magnificent buildings in Indian, which took a toll on trees and forests. The princes in the former states in these colonies are also known to have built massive and comfortable palaces for the colonial masters. The trees metamorphosed into timber logs to support these palaces.

The important thing to notice here is that trees were not only used for aesthetic purposes in these structures. In fact, the trees were already felled for a different reason. But later, this timber was put to use to build these palaces. The aesthetic values attached to such structures came into vogue much later.

Were there any imposing wooden houses built by the pre-colonial kings in India, or did timber form a major part of those magnificent monuments built by the Moguls? Mostly, not. Moreover, the famous London Tower, built in the medieval ages, also does not have any significant use of timber. In India, in and around Rajasthan and Saurashtra, one can see the elaborately carved palaces and edifices built in wood during the late medieval times bordering on the early colonial era. Bihar and Orissa, too, show a similar practice. In Kerala, particularly in the Malabar Coast, one can see houses built of wood. Upon examining the history of land ownership or upon scrutinizing the transition of political regimes, it becomes clear that wherever there was a conflict between the farmers and the rulers, massive treeless fields were created as a result. Later, these felled trees were used for building large homes by intricately carving the wood. Thus, the culture and heritage of this area prospered, founded on the destruction of forests. Probably, the entire history of deforestation, whether during the colonial era or the previously existing feudal rule, is basically the history of injustice meted out to the farmers by their landlords. The merchant class, which supported the feudal landlords, fed the Industrial Revolution with the surviving forests of the world. While in America, as an American academic friend once told me, not a single tree prior to the World War I is found in or around the city of Washington. All of the trees one sees there today were planted in the 20th century. How and why did the indigenous forests disappear?

When I was trying to raise funds for a language research centre that I established during the 1990s, I came in contact with a gentleman who at that point in time was 93 years old. He, Govardhanbhai Patel, still looked energetic and was a well-known philanthropist. He was a generous philanthropist, and upon my request, he made some funds available for my venture. He told me that I should regularly report to him as to the progress of the centre. Accordingly, I would meet with him regularly to update him on the progress. At every meeting, I became familiar with the man's deep devotion to Mahatma Gandhi. Like Gandhi, Govardhanbhai had also gone to Africa in his youth. In the 1930s, he got a job as a supervisor in a warehouse. During the World War II, England had engaged many contractors to clear forests and provide the wood to sustain their war. Govardhanbhai was one of them. After the war was over, the British officer, pleased with his contribution, offered to him a lease of 25 years to some of those forests which spread over thousands of acres. After a few years, a war-ravaged German industrialist offered Govardhanbhai the technology to manufacturing matchsticks. Using these two gifts, he manufactured matchboxes from the trees of those thousands of acres of forest lands. Through the interest he earned from this wealth, Govardhanbhai had been funding innumerable schools, hospitals and other philanthropist projects in Gujarat. He earned a good name as a philanthropist. The forests, however, have vanished. In the three centuries following Columbus reaching America, the American forests disappeared. The British colonials landed in India; Indian forests disappeared. The British and the French landed in Canada and decimated the Canadian forests. Patrick White's Nobel

Prize-winning *Voss*, a novel about the colonialization of Australia, narrates the story of how Australian forests were put to the axe by engaging the criminals banished from England and Germany.

After the devastation of forests in America, Asia and Australia, the colonial imperialists focused on Africa during the mid-20th century. Joseph Conrad has brought this entire annihilation into focus in his novel *The Heart of Darkness*. The British and French deforested major parts of North America, Asia, Australia and Africa. The Spaniards took care of destroying the forests in South America, and the ferocity was no less.

Indentured and slave labour was exploited by the colonizers for turning huge expanses of forests into their "civilized" colonial territories. Did no one ever question the effect of so much of environmental destruction throughout the five continents? Did no one oppose the local feudal regimes exploiting forests, industrialization implicit in the colonial economies and the violent wars generated by European countries devouring the resources of other continents? Didn't the indigenous communities who lived in these jungles and by these jungles, worshipped the forests and preserved them as their sacred homes, resist or protest this assault? The history which holds answers to all these disturbing questions has been hushed up and excluded from any of the recognized history books in academic circulation. Bengali author Mahasweta Devi echoes this heart-wrenching exploitation of the forests in her novels like *Jungle Ke Davedar* (The First claimants of the Forests). This is the kind of history that History should have legitimized, but is mostly recorded in the form of literature that gives the answers to such disquieting questions.

The history I was spoon-fed in school is a distant memory for me now. I studied my history syllabus, which was all about the British laws in India. It did not refer even in passing to the *Criminal Tribes Act* enacted during 1871 declaring the nomadic indigenous communities as "Criminal." In 1872, lists of the "uncouth" tribes of India were made by the colonial administration. However, not a mention was made of this history of exclusion and stigmatization in our history textbooks. If such history is not regularly fed to the Indian children, the glorious movement of forest preservation known as "Ulgulaan," a major revolt by the indigenous against the colonial masters led by Birsa Munda, is all but relegated to obscurity. The splendid "Chipko" movement (movement against forest encroachment) was the result of sheer devotion of the tribal women for their forests. The Narmada Bachao (Save the Narmada River) movement against the building of dams across the Narmada River was fought by tribals living along the river. Industries, mining, railway projects and highway construction are all directly responsible for deforestation around the world. This lopsided progress is imposed on the tribals without any well-informed consent from them and by crushing their protests as anti-national revolts. Their remonstrations for forest preservation have been conveniently erased from all history books. It is therefore not difficult to understand why urban people are normally unaware of the attachment that the indigenous communities have to their forests, hills and rivers.

It is an indisputable fact that the Industrial Revolution, colonialism and capitalistic economics – all important factors constituting the world's modernity – are responsible for the mass deforestation and exploitation of natural resources through mining and relocation of the resources all around the world. The indigenous inhabitants residing for centuries in these forests and hills and along these rivers have been rendered paupers. These three factors were also accountable for disentitling the tribes of their rights on their forests. Yet, when we speak of environmental degradation and a global ecological emergency, a lot of blame for the loss of forest cover is passed on to the indigenous.

The Surat-Howrah railway line was built in India during the colonial times. To provide logs for sleepers for the train tracks, the forests on the border of Maharashtra and Madhya Pradesh were used. The native Korku tribes who lived in some parts of these forests were simply displaced from their homes and their traditional forests were felled. For many decades now, the Korkus have become victims of malnourishment and starvation as the lands given to them in exchange of their original forest land is uncultivable. The continued malnutrition has accentuated the genetic "sickle cell" disease in them, making them fall like parched leaves. This is the condition of tribes, more or less, all over the world today. The indigenous communities have become victims of the greed economic paradigms have legitimized over the last few centuries – greed that nations present to their people as development and progress.

The tribals have been disentitled from their forest homes, hills and rivers, but what have they received in the bargain? They have received river dams, mined hills and the Forest Department! They have received unsympathetic beat guards, and corrupt and greedy forest rangers. In order to get access to land for cultivation, the indigenous communities in India have to pay "fines," a kind of a lease fee or rent. And if the indigenous try to become rebellious, the governments offer to them the promise of "Joint Forest Management (JFM)" – which is normally a promise without much delivery of real controls. The British, French and German colonialists were known to bribe the indigenous communities of America with colourful beads and claim their gold deposits in return. The indigenous communities in India, likewise, have been "gifted" with a few English words like "JFM," "Beat Guards" and "Reserved Forest" in exchange for their land and their natural resources. During the colonial era, the Indian forests became "English," and soon thereafter, they became "alien" to the natives of those forests.

Often, I witness trucks from Madhya Pradesh going towards the urban centres of Gujarat, ferrying cattle, goats and poultry for slaughtering. While watching them, I imagine invisible trucks carrying the rights of ownership of the tribes. The ownership of the forests has been shifted from rural India straight to the bureaucracy of the state in Delhi. The tender touch of the tribals, who have traditionally preserved and looked after the forests, is now being viewed by the regulating authorities as unforgivable. The forests have all but lost the skill of traditional conversation with their children, the indigenous communities. This language of kinship no longer matches with the language of law. Sadly, the forests have no alternative but to learn the definitions declared in the laws and bye-laws enacted in English in order to retain their existence.

Ancient Indian mythology has an interesting story that can sum up the situation: Shiva began rendering many stories to his consort, Parvati. Two of their attendants eavesdropped out of curiosity about these divine stories. As Parvati catches them in the act, she curses them. "You must now on live among humans. They, you shall never recall these stories; and if you do, they will not be in any civilized tongue." They begin their life among humans, but live in a forest. Gradually, however, these stories start returning to their memory, and one of them starts writing them with a rough twig as his pen and the blood of a beast as his ink. Later, when the rush of memory subsides, they go to a king in a city and ask to be gifted for the wonderful stories they have. The king shows no interest in these stories written in blood and in a language that is from the forest. They return heartbroken to the forest and start offering their work to the fire. While burning each story, they read it out aloud to themselves. Birds and beasts gather to listen to these enchanting stories, forget hunger and thirst, and soon enough became emaciated. The king in the city becomes worried, as he does not get a healthy hunt for his meals. He is told that birds and beasts no longer eat in the forest. Finally, the king himself arrives in the forest and manages to snatch the half-burnt manuscript out of the fire. What was saved, as the myth goes, later became the great Sanskrit classic of fiction, the *Katha-sarit-sagar* – the ocean of stories.

The condition of the forests in India today is like that of the two attendants of Shiva in this mythical story. If they speak the traditional idiom of relationships, there is no protection for them. If the forests speak like the indigenous about tenderness, love and affection, there is no one among the regulators to listen to them. If the forests submit themselves to the laws in English, they are readily given a place of pride in the global discourse of the carbon credits counted for further industrial expansion.

The forests of the indigenous and the government laws in English are both in waiting: the forests hope that the laws will take an empathetic view of them; and the laws expect that the indigenous and their forests will integrate "legal literacy" and will willingly give up and set aside their own self-esteem. "The development of the forests" has become a meaningless phrase used only in the loveless documents of the governments of the world. This gloomy vulnerability of the indigenous and of their forests can only be protected when the world learns learn a new language of relation and love. Till that time, the forests will speak those broken words in English, or French or Spanish.

Note

1 This is an excerpt from G.N. Devy's nonfiction work *Vanaprastha* [Entering the Forest], translated from Marathi by Kavita Karnad. Originally published in Marathi as Ganesh Devy, *Vanaprastha* (Pune: Padmagandha, 2008).

3

INDIGENOUS PEOPLES AND THE GREAT LAKES IN NORTH AMERICA

Deborah McGregor

> The Great Lakes were first visited by Europeans over four hundred years ago, and since then, the Lakes' environs have been severely altered. The evidence can be seen everywhere-massive multi-land highways, huge mega-cities, large-scale technological 'improvements' that have altered vast landscapes, mines, power plants, nuclear generating stations, paper mills, steel smelters, water diversion projects-the list goes on and on. In technology's wake we see pollution of the air, water, our land, and even our own bodies.
>
> (Bellfy 2014)

The Great Lakes Ecosystem in North America holds 20% of the fresh water in the world and over 40 million people. The Great Lakes Ecosystem is the largest in-land water transportation system in the world and is regarded as having the best source of fresh water in the world (Freedman and Neuzil 2018). The Great Lakes sustains over 4,000 species of flora and fauna and one of the most ecologically diverse ecosystems on Earth (US-EPA and ECCC 2017a, 2017b: 2).

Long before European colonization and settlement, indigenous nations relied upon and governed the Great Lakes for their long-term sustainability and nationhood. Since colonization, settlement, industrial expansion, capitalism and now globalization, indigenous nations and the Great Lakes face a number of sustainability challenges. Indigenous peoples are impacted in distinct and profound ways due to their thousands of years of relationships with the Great Lakes ecosystem (Perez 2014). This contribution highlights some of these environmental and sustainability challenges from the perspective of indigenous peoples and actions taken to regain decision-making authority and jurisdiction of the Great Lakes. This will be achieved through a discussion of indigenous engagement in Great Lakes governance via the Great Lakes Water Quality Agreement (GLWQA), a bi-national agreement between Canada and the United States to protect the Great Lakes. The purpose of the GLWQA is to restore and maintain the chemical, physical, and biological integrity of the waters of the Great Lakes. To achieve this purpose, the parties agree to maximize their efforts to:

(a) Cooperate and collaborate;
(b) Develop programs, practices and technology necessary for a better understanding of the Great Lakes Basin Ecosystem; and
(c) Eliminate or reduce, to the maximum extent practicable, environmental threats to the waters of the Great Lakes.

The parties, recognizing the inherent natural value of the Great Lakes Basin Ecosystem, are guided by a shared vision of a healthy and prosperous Great Lakes region in which the Waters of the Great Lakes, through sound management, use and enjoyment, will benefit present and future generations of Canadians and Americans.

The parties recognize that it is necessary to take action to resolve existing environmental problems, as well as to anticipate and prevent environmental problems, by implementing measures that are sufficiently protective (ECCC 2013: 5–6).

First signed in 1972, and for four decades afterwards, the GLWQA agreement did not consider the unique experience and perspectives of indigenous nations in the Great Lakes who were thus excluded from governance and decision-making (Cole et al. 1997). Exclusion from participation in agreements, such as the GLWQA, that affect the lives of indigenous peoples is a form of environmental injustice (Whyte 2011). While non-indigenous interests, industry, governments, environmental organizations and others benefit economically from the exploitation of the Great Lakes, indigenous peoples suffered economically, spiritually, culturally, socially and in terms of health from the degradation of the Great Lakes ecosystem. Indigenous peoples not only experience the negative impacts of a degraded Great Lakes ecosystem, they are also impacted politically through exclusion from meaningful decision-making processes. Indigenous peoples have expressed an interest in formally participating in the Great Lakes governance, including the GLWQA and any resulting agreements that flow from it.

In this chapter, I will show how the most recent iteration of the GLWQA (2012) has generated openings for indigenous nations to begin to regain their inherent responsibilities and rights for caring and protecting the Great Lakes, yet still retains features that continue to alienate indigenous nations from exercising their authority and jurisdiction via Great Lakes governance. I will reflect upon a number of initiatives undertaken by the parties (United States and Canada) and indigenous nations to better reflect the distinct legal and governance aspirations of indigenous peoples in the Great Lakes. I will also highlight cross-cutting and specific Great Lakes environmental issues raised by indigenous nations over the years, as well point to future opportunities for respectful collaborations and partnerships.

Indigenous peoples and the Great Lakes

Indigenous interest and concern for the Great Lakes is not new. For thousands of years, the Great Lakes Basin was governed by indigenous internationally (between nations through treaties), regionally (e.g., confederacies) and locally (through clan systems) (Johnston 2005). Many of these aspects exist to this day (King 2014; Whyte et al. 2018). Indigenous governance and autonomy was recognized by the British Crown through the Royal Proclamation of 1763 (Borrows 2007). At that time, there was no boundary between Canada and the US. The boundary set out in the Royal Proclamation was to establish "Indian country," where "Indians" would not be "molested and disturbed." The Royal Proclamation set up the official process for settlement, and from the Crown's point of view, involved the extinguishing of Indian title to the lands coveted by the Europeans powers (Borrows 2007). Many of the treaties negotiated afterwards with indigenous nations sought to remove, relocate or simply erase indigenous presence on the lands/waters to make way for settlement and industrial explosion (Perez 2014).

However, indigenous peoples sought to retain control over their lives, lands and resources through the treaty-making process (RCAP 1996). Deliberate laws, policies and practices were implemented over the past four centuries to eradicate indigenous peoples as distinct peoples and to alienate them from their traditional territories. Such policies formed part of the national government's explicit policy agenda, for example, the Indian Act in Canada and the Indian Removal Act in the United States, and the implementation of the residential/boarding school system in Canada and the United States. Through such processes, indigenous peoples have been disenfranchised from most decision-making on their traditional territories, including the lands and waters in the Great Lakes Basin. In addition to colonizing efforts, industrial use and expansion has greatly impacted indigenous governance and way of life. Indigenous laws, governance and knowledges were replaced with colonial laws and governance structures, effectively excluding indigenous nations from Great Lakes decision-making. These processes undermined indigenous sovereignty in North America, including in the Great Lakes Basin. In this sense, political alienation of indigenous nations from the Great Lakes has roots in a colonial history and agenda, and was thus deliberate.

Despite greatly diminished formal and recognized governance responsibilities in the Great Lakes, contemporary indigenous interests have not waned (Bellfy 2014; Johnston 2005). Indigenous peoples in the Great Lakes Basin continue to rely on the ecosystem for food, water and medicines (COO 2001; Noonan 1999; GLIFWC 2016), and possess treaty rights to exercise such reliance. As such, indigenous people seek involvement in Great Lakes initiatives and continue to engage in their own initiatives. Despite such interest, however, indigenous peoples remain disenfranchised from federal, provincial and state government-led environmental initiatives affecting the basin (Noonan 1999; Ettawageshik 2014).

Contamination of the Great Lakes water, fish and wildlife remains a major concern to indigenous people in the Great Lakes Basin. For example, indigenous nations have long recognized contamination issues as demonstrated in 1990 in the Effects on Aboriginals from the Great Lakes Environment (EAGLE) environmental health contamination research project, initiated to examine the interplay between Aboriginal peoples, their health and contamination in the Great Lakes Ecosystem (COO 2001). Indigenous nations were found to be more vulnerable to the negative impacts of contamination in the Great Lakes than their non-indigenous counterparts. In the United States, treaty organizations such as the Great Lakes Indian Fish and Wildlife Commission (GLIFWC) has been active in relation to the Great Lakes initiatives, including expressing concern over contamination of fish in the Great Lakes. GLIFWC has been an active tribal agency for some time in the Great Lakes Basin in the United States. The GLIFWC is an inter-tribal, co-management agency committed to the implementation of off-reserve treaty rights on behalf of its eleven Ojibwe member tribes. Formed in 1984, it exercises authority specifically delegated by its member tribes. GLIFWC has participated in Great Lakes initiatives and continues to provide leadership for tribes in the United State (GLIFWC 2018).

In Canada, the Anishinabek/Ontario Fisheries Resource Centre (A/OFRC) was established in 1995 to serve as an independent source of information on fisheries assessment, conservation and management, and promotes the value of both Western science and traditional ecological knowledge. Its work is conducted specifically for the Anishinabek Nation (Union of Ontario Indians), which is comprised of 43 First Nations in Ontario, most of which are located in the Great Lakes Basin (UOI 2008). The A/OFRC also completes studies to assess the health of fish in the Great Lakes basin.

Despite the existence of treaty organizations, such as the Great Lakes Indian Fish and Wildlife Commission and the Anishinabek Nation in the Great Lakes Basin, indigenous peoples continue to experience challenges in meaningful participation in the Great Lakes governance. There is no joint indigenous organization that coordinates tribes, First Nations or Métis in Canada and the United States; instead, indigenous nations focus on common challenges and gather when required. For example, in 2004, tribes in the United States and First Nations in Canada gathered over concerns regarding bulk water export and diversions of Great Lakes water and other issues. The nations that gathered signed the Tribal and First Nation Great Lakes Water Accord (2004). Indigenous nations from Canada (Ontario and Quebec) and from the United States (New York, Pennsylvania, Ohio, Indiana, Illinois, Michigan, Wisconsin and Minnesota) signed the Tribal and First Nations Great Lakes Water Accord in November 2004, in Sault Ste. Marie, Michigan. The accord is in response to the Great Lakes Charter – Annex 2001, and asserts their collective interests, concerns, jurisdiction and rights regarding government decisions and management of the Great Lakes Basin waters (Ettawageshik 2014). The United Indian Nations of the Great Lakes met again in Niagara Falls, Ontario, in April 2004 to discuss a path forward to ensure their participation in decision-making involving management of the Great Lakes. The Tribal and First Nations Great Lakes Water Accord states:

> that our rights and sovereignty be respected that any governmental effort to protect and preserve the Waters of the Great Lakes Basin include full participation by Tribes and First Nations, and we also hereby pledge that we share the interests and concerns about the future of the Great Lakes Waters, further pledging to work together with each other and with other governments in the Great Lakes Basin to secure a healthy future for the Great Lakes.
>
> *(Ettawageshik 2014, 204)*

Politically and legally, indigenous nations assert sovereignty and continued rights, responsibilities and duties to protect their territories, including the Great Lakes. Indigenous nations assert they are not subdivisions of colonial governments and are self-determining. Furthermore, a number of First Nations (e.g., Walpole Island, Saugeen/Nawash, Akwesasne) have filed, or intend to file, land claims to the beds of Great Lakes Basin waters. Indigenous nations are increasingly positioning themselves more than ever politically and strategically on water quality and quantity issues (Kempton 2005). The intention is that indigenous participation in the Great Lakes will not be a question of stakeholder involvement, but will transform to constitute governance, rights and sovereignty considerations. These efforts by indigenous nations illustrate the continued exercise of governance, authority and jurisdiction responsibilities along traditional governance lines as well, such as the Six Nations Confederacy through the Haudenosaunee Environmental Task Force (HETF 1999) and the Council of the Three Fires Confederacy (Anishinabek Nation 2018).

In contrast, the main mechanism that is used between Canada and United States to govern the Great Lakes is GLWQA.

The Great Lakes Water Quality Agreement (GLWQA) 2012

Contemporary Great Lakes governance is complex. This complexity is compounded by the numerous jurisdictions involved, including Canada and the United States, with their provinces, states, municipalities and numerous tribes, First Nations and Métis peoples. Great Lakes governance is further compounded by the involvement of numerous non-government actors such as environmental groups, industries and industrial associations, citizen groups, academics, scientists, etc. Not only are there numerous jurisdictions at play (and sometimes at odds with each other), a multitude of non-state actors also have a role in governance. Furthermore, Great Lakes governance is evolving. Effective governance systems evolve over time as different challenges are faced. Great Lakes governance is no exception.

Water and water rights issues have been ongoing between Canada and the United States for over a century. To help resolves disputes between the two countries, the Boundary Waters Treaty (BWT) was signed in 1909 to prevent and resolve disputes over the use of the shared waters and to settle other transboundary issues. To assist the two countries in their work, the International Joint

Commission (IJC) was established to carry out the provisions of the treaty. The BWT serves as the legal and institutional foundation for contemporary Great Lakes governance.

In response to serious water quality concerns in the Great Lakes, coupled with the recognition of the legitimacy of the emerging environmental movement, both Canada and the United States adopted environmental policies and plans to address environmental issues for many regions including the Great Lakes. Canada and the United States agreed that action was needed to address Great Lakes environmental issues. First signed in 1972, the current Great Lakes Water Quality Agreement (2012) reflects the long-standing history of cooperation between Canada and the United States to protect the Great Lakes.

Various iterations of the Great Lakes Water Quality Agreement over the past four decades have established a shared vision and common objectives and commitments to science, governance and action that will help to restore and protect Great Lakes water quality and the health of the ecosystem (ECCC 2013). Unfortunately, these historical international and domestic agreements did not consider the long-standing governance of indigenous nations, including existing treaties, rights and knowledge until relatively recently. It must be noted that indigenous nations have governed the Great Lakes through their own laws, treaties, agreements, governance structures, knowledge and values for thousands of years. Disruption to indigenous nations' authority and jurisdiction is centuries old, which is relatively recent compared to the thousands of years of indigenous relationships and responsibilities in the Great Lakes. However, unlike previous iterations of the GLWQA, the 2012 version includes provisions for indigenous engagement and consideration of traditional ecological knowledge. It has taken 40 years for the unique aspirations and contributions of indigenous peoples to be recognized by Canada and the United States and included in the GLWQA. The GLWQA includes the following provisions:

> RECOGNIZING that, while the Parties are responsible for decision-making under this Agreement, the involvement and participation of State and Provincial Governments, Tribal Governments, First Nations, Métis, Municipal Governments, watershed management agencies, local public agencies, and the Public are essential to achieve the objectives of this Agreement.
>
> *(4)*

It also recognizes indigenous knowledges in the Approaches and Principles section of the agreement:

> 4 l) science-based management – implementing management decisions, policies and programs that are based on best available science, research and knowledge, as well as traditional ecological knowledge, when available;
>
> *(7)*

For the remainder of the agreement, tribal governments, First Nations and Métis are mentioned along with state and provincial governments, and other agencies to help deliver the objectives of the agreement and implement them.

In terms of Great Lakes governance, the GLWQA established the Great Lakes Executive Committee (GLEC) co-chaired by the "parties," which meets twice a year:

> the Parties shall co-chair the Great Lakes Executive Committee and invite representatives from Federal Governments, State and Provincial Governments, Tribal Governments, First Nations, Métis, Municipal Governments, watershed management agencies, and other local public agencies
> *(12)*

The GLWQA includes a Great Lakes Public Forum to report to the public on the status of the Great Lakes and actions taken to address identified priorities. The International Joint Commission (IJC) is also required to engage with indigenous governments. Engagement with indigenous governments is also mentioned throughout the ten annexes of the GLWQA. Despite the inroads made in this iteration of the GLWQA for indigenous engagement and recognition of indigenous governments and traditional knowledge, limitations are evident. The main purpose for the involvement of indigenous governments is to deliver on the priorities and issues identified by the parties (the Canadian and American governments), and to recognize the rights of indigenous nations to exercise their own jurisdiction to protect the Great Lakes waters as outlined in the various treaties and the more recent Tribal/First Nation accord.

Furthermore, from an indigenous perspective, any Great Lake governance initiative has to recognize that indigenous nations reveal a higher degree of vulnerability to health effects from Great Lakes contaminants than occurs in non-indigenous people. This is due to historical indigenous occupation of the area and reliance on (now contaminated) fish and wildlife. Respiratory ailments due to poor air quality and various impacts from poor water quality, especially drinking water, are more prevalent in indigenous communities in the Great Lakes than in their non-indigenous counterparts (COO 2001; UOI 2009). These health and environmental injustice issues require resolution to support healthy indigenous nations (LaDuke 1999).

Another critical aspect not considered in the GLWQA (2012) is the historical and ongoing process of colonization that has produced the current context of indigenous alienation in every respect of Great Lakes governance. As noted earlier, historical colonial laws, policies and practices were deliberate in their undermining of indigenous nations. Until these underlying political, legal and governance issues are addressed, indigenous nations will continue to suffer.

Over the centuries, indigenous nations have continued to advocate for their inclusion in Great Lakes governance. As part of their advocacy, they have also participated in a number of Great Lakes fora and have identified environmental challenges of concern to them. These concerns are summarized in the next section.

Indigenous Great Lakes environmental issues

Despite the exclusion and lack of recognition of indigenous nations from Great Lakes governance, there have been various engagement initiatives involving indigenous peoples in the Great Lakes Basin over the past two decades. Some initiatives include (but are not limited to): participation and presentations to International Joint Commission (IJC) meetings; State of the Great Lakes Ecosystem conferences (SOLEC); and Great Lake Water Quality Agreement (GLWQA) and Great Lake program renewal discussion/dialogue sessions. Indigenous communities have also been involved in federal, provincial and state program-level environmental conservation and protection efforts. Indigenous advocacy and engagement in these initiatives reveals consistent themes in historical, current and emerging environmental issues and concerns from an indigenous perspective.

Indigenous nations have identified a number of environmental concerns relating to the Great Lakes, their homeland for thousands of years. Indigenous nations have also raised a number of concerns relating to the exercise of their sovereignty, self-determination, treaty rights and governance in relation to the Great Lakes. The following section provides a summary of key cross-cutting issues raised by indigenous peoples followed by more specific environmental concerns identified in Table 3.1.

Indigenous governance

Indigenous governance is not new. For thousands of years, the Great Lakes Basin was managed through indigenous governing structures – many aspects of which still exist to this day. Indigenous governance and autonomy was recognized by the Crown (initially the British, and later Canadian and American) through the Royal Proclamation and later treaty-making processes (RCAP 1996; Rashidi 2014; GLIFWC 2018).

There are also treaties and governance agreements that current governments disregard as part of the treaty relationships, for example, the pre-colonial Dish with One Spoon treaty between the Haudsonsaunee and Anishinaabek is a treaty that agrees to share resources among nations, and the Two Row Wampum treaty embodying the principles of peace, friendship and respect. These types of agreements contributed to the "governance" of the Great Lakes for thousands of years before contact (King 2014; Whyte et al. 2018). Governance of the Great Lakes did not begin with the Boundaries Water Treaty (1909) or the Great Lakes Water Quality Agreement (GLWQA). There is a long history of governance based on different political philosophies and legal systems in the Great Lakes Basin involving different indigenous nations and with Europeans who arrived to settle (King 2014).

TABLE 3.1 Great Lakes environmental issues

Great Lakes Issues
Industrial Pollution Industry has been polluting the Great Lakes for over a century. Indigenous nations are dismayed that it is legal to dump chemicals and other contaminants into the Great Lakes. Atmospheric pollution also deposits contamination in the Great Lakes.
Economic Impacts of Great Lakes Degradation The economic livelihood of indigenous nations has been impacted by decreased fisheries stocks (dissemination of fisheries). Other economic losses occurred due to environmental degradation.
Compensation First Nations raised the issue that environmental contamination of the Great Lakes has impacted their health and livelihood in negative ways. First Nations indicated that the "Crown" has to consider how First Nations will be compensated for damages inflicted upon them from environmental degradation of the Great Lakes. Compensation for environmental damage has not been given due consideration in Great Lakes discussions. How will First Nations be compensated for the injustice, health and economic impacts they have endured?
Water Governance and Management First Nations in the Great Lakes Basin continue to experience boil water advisories. The First Nations wish for engaged water monitoring in their communities and in activities that occur off-reserve in their traditional territories that impact water quality.
Water Quality Water quality for human life and wildlife is regarded as equally important. There are additional concerns around bottled water, including safety and waste management. Other water quality issues include: nutrient loading, leaching of pesticides and the presence of pharmaceuticals in the water. Indigenous nations are concerned about water quality standards (not stringent enough to protect water for human and all life). Some indigenous nations still do not have access to treated water. Many indigenous nations' community members continue to rely on groundwater sources, wells, etc. (not treated).
Environmental Health: Toxins First Nations have experienced poor health due to decreased water quality in the Great Lakes, including the community of Aamjiwnaang First Nations low ratio of male to female births. First Nations report higher cancer levels attributed to environmental contamination (Mackenize et al. 2005). Chronic health conditions such as diabetes and autoimmune disorders have been attributed to contamination. "Direct clinical health effects include a marked increase in cancer, birth defects, diabetes, reproductive and hormone-based disorders, and immunological-based disorders" (Cole et al. 2007, 11).
Dredging of Contaminated Sediment: Stirs Up All Kinds of Contaminants Open dumping in waters is a problem for indigenous nations. Clean-up may cause more harm, as contaminants will be released in to the waters again.

Great Lakes Issues

Environmental Health: Contaminated Foods
The EAGLE Project's main finding points to unacceptable levels of contamination in the waters and on the lands that support the lives of First Nations people in the Great Lakes Basin. There is clear evidence that Great Lakes First Nations consume large quantities of fish, significantly more than the general Canadian population. Persistent toxic substances (PTSs) such as methylmercury and PCBs are present in the fish consumed by First Nations.

Aquaculture: Fish Farms and Stocking
Indigenous nations are concerned that aquaculture contaminates aquatic life (other fish, etc.). Aquaculture operations involve nutrient loading in the waters from fish food and waste. Farmed fish also carry diseases that can spread to native species. Stocking of lakes and rivers with non-indigenous species (e.g., salmon, splake) is destroying stocks of native fish. Stocking is about appeasing the powerful sport fishing lobby, which has no understanding of the biological or ecosystem impacts.

Sharing Information on Fish Consumption Sources of Contamination
Indigenous nations do not have information on the contamination of fish they consume (source of pollution). Fish consumption guidelines need to reflect the consumption and values of First Nations. The EAGLE project showed that First Nations consume more fish than non-natives in the Great Lakes. First Nations are not able to make informed decisions about their health if they do not have up-to-date, correct information.

Raw Sewage Discharge
Indigenous nations have consistently pointed out that tourists in the Great Lakes dump their raw sewage into the water. Tourist operators and yacht owners do the same.

Municipal Waste Water
Indigenous nations continue to express concern about the waste waters (sewage, etc.) released by cities and municipalities.

Unsustainable Water Use and Water Permits
Indigenous nations expressed concern that water use is unsustainable by industry, agriculture and municipalities, as well by individuals (household use). Governments have also issued permits for industry to take water from groundwater sources (without consent or engagement of indigenous peoples).

Habitat Protection Around the Great Lakes
Protecting habitats is an important part of protecting the Great Lakes. There is a need to better understand how what happens on the land at the landscape level affects what happens to the Lakes. Habitat protection has to occur on a broad scale (ecosystem or landscape level) as well as the community level.

Aquatic Invasive Species
Invasive species (e.g., zebra mussels, goby) and, more recently, Asian carp species remain a critical concern to indigenous nations communities. Invasive species out-compete native species for food and habitat.

(*Continued*)

TABLE 3.1 (Continued)

Great Lakes Issues
Ship-Source Pollution The Great Lakes are contaminated on a continual basis from ships via ballast water from ocean vessels, toxins and dangerous substances shipped through the Great Lakes.
Lack of Monitoring and Enforcement There are environmental protection laws that apply to the Great Lakes, but they are not enforced. Some activities need to be "banned" outright, not just "regulated (e.g., Ballast water). Industry needs to be monitoring more closely.
Climate Change Extremes in temperature are noticeable, as are severe storm events (flooding). There is a general warming. Ice fishing starts later and duck hunting can continue until Christmas. "Causes damage to sugarbush and fruit bushes. Traditional medicine people cannot find natural ingredients to prepare medicines" Noonan 1999, 19). Some migratory birds are staying year-round. "Rare plant species have been damaged in flash-flooding along rivers and creeks in the last couple of years" and "Reduced water levels result in water freezing of the bottom of the lake bed ruining spawning grounds" (Noonan 1999, 19). Climate change will increase water temperatures and affect flow rates of streams and rivers. Climate change and associated severe weather patterns will impact the indigenous community's infrastructure (flooding, growth of mould).
Culture, Heritage and Future Generations Protection of the Heritage and Culture value of the Great Lakes is of utmost importance. Water is regarded as a sacred gift and First Nations have an obligation to care for the waters. Waters are important in ceremonies and healing, as shown by the Great Lakes Mother Earth Walkers (McGregor 2015). A degraded environment impacts cultural survival. It has also been noted that indigenous names of lakes and waters have been changed into English/French, thus losing an important cultural heritage. This is related to a loss of traditional culture and language (as language is not used to describe places anymore). Indigenous nations remain concerned about the water and environment they will inherit. The Women's Waters Walkers have observed that First Nations peoples are losing their connection to water (especially when water is degraded).
Contaminated Sites There are many contaminated sites around the Great Lakes (traditional territory) and in indigenous nations that pose a concern. Contaminated sites impact water quality via contamination of the groundwater or runoff from surface water.
Lack of Research on Issues of Concern to Indigenous Peoples Indigenous nations have called for research on various issues (invasive species, species at risk, health studies, biodiversity, base line studies, fish contamination studies, etc.). There needs to be more partnerships with academics and others to conduct independent research.
Lack of Coordination Even Among FN/Tribes to Strategize Around "Water" Issues Tribes and First Nations need to have their own conferences and meetings to assess the Great Lakes Basin Ecosystem, conduct their own research and develop their own priorities.

Great Lakes Issues

Shoreline Development/Shoreline Erosion
Increased boat and vessel traffic has impacted the shoreline (erosion) of some First Nations.

Environnemental Justice
Cole et al. write "Indigenous communities are disproportionately impacted and are becoming 'human indicator species' to toxic contamination" (17). Indigenous communities now often suffer from "myriad toxic legacies placed upon them by centuries of (first) colonial oppression and (ongoing) post-colonial exclusion from the decisions and benefits of natural resource and industrial development which have occurred, in many cases literally, in their backyards." (30)

Historical Environmental Burdens
Indigenous nations have suffered as a result of environmental contamination in the Great Lakes. Indigenous nations have relied on the Great Lakes for their cultural, spiritual, social and economic livelihood. The inability to eat the fish, drink the water, and use the medicines, for example, has had detrimental impacts on First Nations people and communities at the individual, family, community and nationhood levels. Contamination of the Great Lakes had led to health issues in indigenous nations communities, both directly or indirectly. Due to four centuries of environmental damage to the Great Lakes, millions of dollars in damage to communities has occurred. There is little funding to deal with historical damage and current damage, let alone future challenges. The historical burden affects current efforts at sustainable economies in First Nations communities and in their territories. Indigenous nations' capacity to engage in "repairing" the damage is limited by lack of capacity (funding, training, etc.).

Near-Shore Activities
First Nations report that chemical spraying occurs along the shoreline to manage weeds (municipalities).
Illegal dumping around waters (contaminants leak into the water) at dump sites.

Nuclear Waste management
Indigenous people remain concerned about the transport of nuclear waste (high-level nuclear waste) across the Great Lakes, as well as the siting of the deep geological repository.

Water Quantity
Great Lakes water diversions and schemes are constantly talked about as the pressure for water increases. Water quantity issues remain a primary concern of First Nations, including privatization, bulk water exports, etc. Climate change is expected to cause further water quantity challenges (decreasing water levels).

Source: McGregor, D & McGregor, H. 2016. [unpublished report]. Environmental Scan: First Nation Great Lakes Concerns. Prepared for the Canada-Ontario Agreement on Great Lakes Water Quality and Ecosystem Health, Annual First Nations Meeting with COA Executives. Brantford, Ontario, p. 24.

Indigenous and treaty rights

The lack of recognition of indigenous and treaty rights is a consistent theme raised by indigenous nations over the decades. Although treaties have been negotiated with the Crown, these rights are ignored or violated on a consistent and sustained basis. Furthermore, such rights also relate to the government's duty – federal, state and provincial – to consult and accommodate indigenous nations (Ettawageshik 2014; COO 2008). Furthermore, the United Nations Declaration on the Rights of Indigenous Peoples (UNDRIP), supported by both Canada and the United States in principle, has not found any expression in Great Lakes governance to date. Indigenous rights are not officially recognized in the GLWQA and thus remain beyond the scope of discussion and dialogue regarding the Great Lakes, despite the advocacy of indigenous nations (Whyte et al. 2018).

Environmental health

The Effects on Aboriginals in the Great Lakes Environment (EAGLE) project, a university and First Nations community-based environmental health study that spanned a decade, found that First Nations are far more vulnerable to a degraded and contaminated Great Lakes environment than other peoples. First Nations are one of the most vulnerable populations in the Great Lakes. Determinants of health such as low income, poor-quality housing and exposure to pollutants means indigenous nations are considered a vulnerable population for exposure to contaminants and other environmental hazards. Indigenous nations have been marginalized in Great Lakes governance, but bear "disproportionate impacts of these pollutants compared to other populations" (Cole et al. 1997, 19). This vulnerability is especially apparent for indigenous nations communities living in and around the Great Lakes Basin Ecosystem. The average indigenous person has poorer health outcomes (Tobias and Richmond 2014). The Anishinabek Nation report "Through the Eyes of a Child: First Nations Children's Environmental Health" reports on indigenous elders and children who are the most vulnerable and are experiencing the ill effects of an unhealthy environment (UOI 2009, 4).

Capacity building

Indigenous nations wish to engage in government-established Great Lakes initiatives. However, the capacity to do so in a sustained fashion is hampered by a lack of capacity to become engaged. First Nations lack environmental units in their communities, which results in inconsistent participation in important meetings. Lack of indigenous nation participation results in a lack of historical knowledge or corporate memory of various initiatives and achievements. These issues have been identified for decades – more specifically the lack of environmental capacity (COO 2001; Noonan 1999). This presents problems with continuous, sustained engagement in Great Lakes-related initiatives. This means that although indigenous nations desire and have a profound responsibility to care for the Great Lakes, they do not have the capacity to do so.

Capacity building is further undermined by historical and ongoing colonialism that continues to marginalize indigenous peoples from Great Lakes efforts.

Indigenous/traditional ecological knowledge

The call for the recognition of Traditional Ecological Knowledge (TEK) by indigenous nations has been articulated for over two decades. In 1993, the Resolution Presented by the Native Caucus convened during the biennial meeting of the IJC in 1993 in Windsor, Ontario states: "The Scientific community must recognize and acknowledge the critical and essential value of indigenous ecological knowledge" (Cole et al. 1997, 4). The most recent GLWQA recognizes TEK along with Western science as part of its principles. Indigenous/traditional knowledge must be given equal weight and respect to Western science in Great Lakes deliberations (McGregor 2014b). Indigenous nations have called for the value of indigenous/traditional knowledge to be respected along with science in Great Lakes governance. This continues to be a challenge in practice. For example, TEK studies are not funded in the same way scientific studies are in the Great Lakes. TEK does not inform priority research questions in the same way as scientific studies do. There is a long way to go in terms of recognition of TEK in Great Lakes work on the ground.

In summary, indigenous nations have asserted that they have much to contribute to the contemporary sustainability of the Great Lakes. In fact, the Great Lakes Ecosystem was governed by indigenous nations for thousands of years prior to (and after) contact with Europeans, and those indigenous nations developed sophisticated knowledge systems to ensure sustained and reciprocal relationships with the Great Lakes. Indigenous people's involvement in the Great Lakes is not new. However, due to devastating historical and ongoing colonial relationships, indigenous nations are among the most economically impoverished peoples living in the Great Lakes Basin (Cole et al. 2007; Griffin and McClenaghan 2011).

The next section provides a summary of identified environmental issues identified by indigenous nations.

Specific environmental issues

Table 3.1 that follows conveys historical, current and emerging Great Lakes environmental issues identified by indigenous nations and peoples.

Emerging trends and considerations

Indigenous nations asserted that their long-standing (and ignored) legal orders must be respected (Borrows 2010; Craft 2014). These legal orders provide the guidance necessary to live sustainably with the Great Lakes (COO 2011).

Although these legal orders are generally not codified, the practices continue to be followed in many indigenous communities. A critical aspect of asserting indigenous legal orders is also to recognize the indigenous governance structures and processes. Indigenous nations governed the Great Lakes for thousands of years and continue in their efforts to do so (King 2014). For example, the Chiefs in Ontario created the Water Declaration of the Anishinaabek, Mushkegowuk and Onkwehonwe to guide First Nations' response to broader water and Great Lakes initiatives (COO 2008).

Indigenous nations have their own laws, treaties, knowledge and governance systems that are not given respect or equal weight to Western laws and knowledge. Indigenous nations have been governing themselves in the Great Lakes region for thousands of years, while at the same time developing the knowledge systems which supported them to do so (Whyte et al. 2018).

Indigenous nations are seeking to reclaim their inherent responsibilities to protect the Great Lakes through the exercise of their responsibilities, relying on their own knowledge systems. Indigenous nations have also expressed concern that the current Canadian and American political/legal framework for protecting the Great Lakes has failed and that attention and respect must be given to indigenous legal orders and traditions. Indigenous legal orders supported sustainability and peace in the Great Lakes for thousands of years.

The assertion of indigenous sovereignty and self-determination does not mean indigenous nations wish to exclude others from Great Lakes governance. In Whyte et al. (2018), a comprehensive study was undertaken of successful indigenous-non-indigenous environmental collaborations and partnerships in the Great Lakes, and recognition of self-determination was identified as a key indicator of success. Successful collaborators seek to formalize their relationships with indigenous peoples through agreements and memoranda of understandings to formally recognize and respect indigenous goals and aspirations.

Conclusions

Despite the inroads made over the past four decades by Canada and the United States to include indigenous nations in the Great Lakes governance through the GLWQA, indigenous peoples are not interested in just informing decision-making, they wish to be part of decision-making from its inception. They are simply interested in implementing the goals and objectives of the GLWQA, but also wish to set direction and priorities for protection of the Great Lakes. Indigenous nations have pointed out that they wish to govern and manage Great Lakes issues in their communities and traditional territories, rather than just providing "input" into other management and governance structures (Ettawageshik 2014). Recognizing again that collaborations, co-governance and sharing jurisdiction with others is necessary to protect, restore and conserve the Great Lakes in the contemporary context, indigenous nations, including laws, governance and knowledge, are still not given respect in the GLWQA.

As noted earlier, the governments of Canada and the United States have failed to recognize the United Nations Declaration on the Rights of Indigenous Peoples in the GLWQA. Recognition by the GLWQA would advance the rights and responsibilities of indigenous nations in Great Lakes sustainability efforts. By no means will UNDRIP address all issues confronting indigenous peoples, but it does offer a framework for advancing self-determination in various contexts, including the Great Lakes. A core aspect of realizing self-determination is maintaining relationships with one's territories and fulfilling responsibilities to the land and future generations. Article 25 of UNDRIP states:

> Indigenous peoples have the right to maintain and strengthen their distinctive spiritual relationship with their traditionally owned or otherwise occupied and used lands, territories, waters and coastal seas and other resources and to uphold their responsibilities to future generations in this regard.
> *(UNGA 2007)*

Indigenous peoples' relationships to environment, territories, lands, waters and the natural world are often framed in terms of responsibilities enacted to ensure sustainability. Despite colonial disruption to the rights, obligations, duties and responsibilities to the Great Lakes, grassroots and community-based initiatives have continued. One important example is the Great Lakes Mother Earth Walks led by Grandmother Josephine Mandamin. The Mother Earth Water Walks were initiated in 2003 with the goal of raising awareness of the sacred connection between people, especially women, and the waters. This action was taken in response to the decades and even centuries of rising pollution levels in the Great Lakes and elsewhere, and the increasing need for people everywhere to take action to protect the waters by renewing their responsibilities to the Great Lakes. In order to achieve this, the walkers, set a goal of walking the perimeter of each of the Great Lakes every year. This movement has generated walks and a water protection movement internationally as well (McGregor 2015). Ultimately, indigenous nations wish to realize their inherent responsibilities to restore, protect and relate to the Great Lakes.

The 2012 GLWQA has made significant inroads in recognizing the importance of indigenous nations in Great Lakes governance, but only on the terms of the Canadian and American governments. The Canadian and American governments have failed to consider the goals, aspirations and priorities of indigenous nations and how the GLWQA can support indigenous self-determination.

Bibliography

Anishinabek Nation. 2018. 'About Us', www.anishinabek.ca/about-us/. (accessed on 28 June 2018).

Bellfy, Phil. (ed.). 2014. *Honor the Earth: Great Lakes Indigenous Response to Environmental Crisis*. Zibi Press.

Borrows, John. 1997. 'Wampum at Niagara: The Royal Proclamation, Canadian Legal History, and Self-Government'. In Asche, M. (ed.). *Aboriginal and Treaty Rights in Canada: Essays on Law, Equity, and Respect for Difference*, 155–172. Toronto: University of Toronto Press.

Borrows, John. 2010. *Canada's Indigenous Constitution*. Toronto: University of Toronto Press.

Chiefs in Ontario (COO). 2008. 'Water Declaration of the Anishinaabek, Mushkegowuk, and Onkwehonw in Ontario', http://chiefs-of-ontario.org/Assets/COO%20water%20declaration%20revised%20march%202010.pdf (accessed on 28 June 2018).

Chiefs of Ontario (COO). 2001. 'Effects on Aboriginals in the Great Lakes (EAGLE) Project Health Survey'. Toronto: Chiefs of Ontario.

Chiefs of Ontario (COO). 2011. 'Honouring the Waters Indigenous Water Forum'. Toronto: Chiefs of Ontario.

Cole, Maxine, J. Warledo, T. Goldtooth, R. Peters, and P. Skye. 2007. 'Environmental Injustice in the Great Lakes Water Quality Agreement'. A presentation to the International Joint Commission, November 2nd, Niagara Falls, Ontario). EAGLE Project, Indigenous Environmental Network and Great Lakes United.

Craft, Aimée. 2014. *Anishinaabe Nibi Inaakonigewin Report: Reflecting the Water Laws Research Gathering*. Winnipeg: University of Manitoba Human Rights Research (CHRR) and the Public Interest Law Centre (PILC), http://static1.squarespace.com/static/54ade7ebe4b07588aa079c94/t/54ec082ee4b01dbc251c6069/1424754734413/Anissinaabe-Water-Law.pdf (accessed on 28 June 2018).

ECCC. 2013. 'A Renewed Commitment to Action: The 2012 Great Lakes Water Quality Agreement', www.ec.gc.ca/grandslacs-greatlakes/default.asp?lang=En&n=B274CBC1-1 (accessed on 28 June 2018).

Ettawageshik, Frank. 2014. 'The Tribal and First Nation Great Lakes Water Accord'. In Bellfry, P. (ed.). *Honor the Earth: Great Lakes Indigenous Response to Environmental Crisis*, 203–208. Zibi Press.

Freedman, Eric and Mark Neuzil. 2018. *Biodiversity, Conservation, and Environmental Management in the Great Lakes*. New York: Routledge.

GLIFWC. 2016. 'Enhancing GLIFWC Capacity to Protect and Restore Lake Superior'. Odanah, WI, www.glifwc.org/publications/pdf/GLRI.pdf (accessed on 28 June 2018).

GLIFWC. 2018. 'Great Lakes Indian Fish and Wildlife Commission'. Odanha, WI, www.glifwc.org/ (accessed on 28 June 2018).

'Great Lakes Water Quality Agreement'. 2012. www.ec.gc.ca/grandslacs-greatlakes/A1C62826-72BE-40DB-A545-65AD6FCEAE92/1094_Canada-USA%20GLWQA%20_e.pdf (accessed on 28 June 2018).

Griffin, Renee and Theresa McClenaghan. 2011. 'Making the Links: Pollution, Poverty, and Environmental Justice'. *Great Lakes Research Review*. 8: 28–33.

Haudenosaunee Environmental Task Force (HETF). 1999. 'The Words That Come Before all Else: Environmental Philosophies of the Haudenosaunee'. Native North American Travelling College.

International Joint Commission (IJC). 2018. 'Great Lakes Water Quality Agreement 2012', www.ijc.org/en_/Great_Lakes_Water_Quality (accessed on 28 June 2018).

Johnston, Darlene. 2005. 'Connecting People to Place: Great Lakes Aboriginal History in Cultural Context'. Prepared for the Ipperwash Commission of Inquiry, www.attorneygeneral.jus.gov.on.ca/inquiries/ipperwash/transcripts/pdf/P1_Tab_1.pdf (accessed on 28 June 2018).

Kempton, Kate. 2005. 'Bridge over Troubled Waters: Canadian Law on Aboriginal and Treaty "Water" Rights, and the Great Lakes Annex'. A paper prepared for the Great Lakes Indigenous Peoples Network, Niagara Falls, Ontario, www.oktlaw.com/pub07.htm (accessed on 28 June 2018).

King, Joyce. 2014. 'Haudenosaunee Position Paper on the Great Lakes'. In Bellfry, P. (ed.). *Honor the Earth: Great Lakes Indigenous Response to Environmental Crisis*, 209–236. Zibi Press.

LaDuke, Winona. 1999. *All Our Relations: Native Struggles for Land and Life*. Cambridge, MA: South End Press.

Mackenize, Constanze A., Ada Lockridge, and Margaret Keith. 2005. 'Declining Sex Ratio in a First Nation Community'. *Environmental Health Perspectives*. 113(10): 1295–1298.

McGregor, Deborah. 2014a. 'First Nation, Traditional Ecological Knowledge and the State of the Great Lakes Ecosystem Conference'. In Bellfry, P. (ed.). *Honor the Earth: Great Lakes Indigenous Response to Environmental Crisis*, 107–125. Zibi Press.

McGregor, Deborah. 2014b. 'Traditional Knowledge and Water Governance: The Ethic of Responsibility'. *Journal of AlterNATIVE: An International Journal of Indigenous Peoples*. Special Issue: Indigenous Knowledges Impacting the Environment. 10(5): 493–507.

McGregor, Deborah. 2015. 'Indigenous Women, Water Justice and Zaagidowin (Love)'. *Journal of Canadian Woman Studies*. 30(2/3): 71–78.

Noonan, Ann. 1999. 'Great Lakes Action Plan: An Aboriginal Perspective'. Toronto, ON: Environment Canada.

Perez, Scott. 2014. 'Native Americans in the Great Lakes/St. Lawrence River Basin: Treaties, Laws and Environmental Issues'. In Bellfry, P. (ed.). *Honor the Earth: Great Lakes Indigenous Response to Environmental Crisis*, 173–182. Zibi Press.

Rashidi, M. 2014. 'Indigenous Peoples and the Great Lakes'. Toronto, ON: Great Lakes Policy Research Network. Ryerson University.

Royal Commission on Aboriginal Peoples (RCAP). 1996. *People to People, Nation to Nation: Highlights from the Report of the Royal Commission on Aboriginal Peoples*. Ottawa: Minister of Supply and Services.

Tobias, Joshua and Chantelle Richmond. 2014. 'That Land Means Everything to us as Anishinaabe: Environmental Dispossession and Resilience on the North Shore of Lake Superior'. *Health and Place*. 29: 26–33.

'Treaty Between the United States and Great Britain Relating to Boundary Waters, and Questions Arising Between the United States and Canada', http://ijc.org/files/tinymce/uploaded/Boundary%20Waters%20Treaty%20of%201909_3.pdf (accessed on 28 June 2018).

UNGA. 2007. 'United Nation Declaration on the Rights of Indigenous Peoples'. United Nations General Assembly.

Union of Ontario Indians (UOI). 2008. 'Anishinabek Traditional Knowledge and Water Policy Report'. Anishinabek Ontario Resource Management Council. AORMC Water Working Group Lands and Resources Department. Union of Ontario Indians. Nippising First Nation. Ontario.

Union of Ontario Indians (UOI). 2009. 'Through the Eyes of a Child: First Nation's Children's Environmental Health'. Anishinabek Nation Health Secretariat. Union of Ontario Indians. Nippising First Nation. Ontario.

US-EPA and ECCC. 2017a. 'State of the Great Lakes: An Overview of the Status and Trends of the Great Lakes Ecosystem', https://binational.net/2017/06/19/sogl-edgl-2017/ (accessed on 28 June 2018).

US-EPA and ECCC. 2017b. 'State of the Great Lakes Highlights Report: An Overview of the Status and Trends of the Great Lakes Ecosystem', https://binational.net/2017/06/19/sogl-edgl-2017/ (accessed on 28 June 2018).

Whyte, Kyle. 2011. 'The Recognition Dimensions of Environmental Justice in Indian Country'. *Environmental Justice*. 4(4): 199–205.

Whyte, Kyle et al. 2018. 'Indigenous Principles for Successful Cooperation'. In Freedman, E. and Neuzil, M. (eds.). *Biodiversity, Conservation and Environmental Management in the Great Lakes Basin*, 182–194. New York: Routledge.

4

INDIGENOUS ART, RESILIENCE AND CLIMATE CHANGE

Australia, Aotearoa New Zealand and Samoa

Tracey Benson, Lee Joachim, Huhana Smith, Penny Allan, Martin Bryant, Tamasailau Suaalii-Sauni, Penehuro Fatu Lefale and Charles Dawson

The storied flows of river, wetland and ocean; the transmission of indigenous knowledge through treasured languages, species and places: indigenous Australian, Pasifika (specifically Samoan) and *Māori* voices gather in this chapter. Each voice is different, as befits both the vast geographical area and varied homeplace stories. Biocultural heritage and collaborative action bind communities and their work.

Cultural flows of history and water coalesce here. Again and again, they affirm the power of indigenous traditions and communities' innovative adaptation to the ongoing violence of (neo)colonial "structural" impositions (Referendum Council).

§

Indigenous Australian songlines, river worlds and realities

Tracey Benson and Lee Joachim

One of the biggest challenges in writing this section of the chapter has been how to address the huge geographical region of Australia and the Torres Strait and diversity of cultures within such a tight word limit.

Gordon De Brouwer, the Secretary of the Department of the Environment and Energy, summed up Australia as being "megadiverse," explaining that:

> Only 17 countries have been identified as megadiverse and Australia is one of them. Our unique environment is made up of many different landscapes and ecosystems – from forests and alpine regions in the south east, tropical rainforests and savannas in the north, deserts in central Australia, and

> wetlands across the country. And Australia is home to more than 7 per cent of the world's total number of species of plants and animals, with about 80 per cent of our mammals and 90 per cent of our trees, ferns and shrubs found nowhere else in the world.
>
> *(2)*

Not only are there a diversity of landscapes, there is also a diversity of cultures that are connected to these places. There are hundreds of First Nation communities in Australia, and so it is impractical to acknowledge all of the important work that is happening across the country's eight climatic zones. It also needs to be stated that this research has been compiled with great respect to the countries which are discussed, and their elders, past, present and future.

The authors of this section, Lee Joachim and Tracey Benson, work together on a project titled "Way of the Turtle," which focuses on well-being and country. Lee is a *Yorta Yorta* elder, researcher and CEO of the Rumbalara Health Service. Tracey was born in Ningi Ningi Country (Redcliffe, QLD) of mixed migrant heritage and works across the arts, education and the public sector. Lee and Tracey's work together focuses on knowledge sharing and empowerment through connectedness to place and story. Lee is also a co-author on a number of the citations, as well as being involved in the GIS community mapping project discussed. The co-authors have consulted with a number of other indigenous elders from the places discussed – Brendan Kennedy Tati Tati Latji Latji Mutti Mutti Weki Weki Wadi Wadi Yitha Yitha Nari Nari, Traditional Owner First Nations of the Murray River, Murrumbidgee River, Lachlan River, Wakool River and Edward River and *Ngunawal*[1] Elder Tyronne Bell.

We explore some of the river countries of Eastern Australia as a way of mapping and layering knowledge systems used to understand the river and to posit rivers systems and fresh water as a critical part of adapting to climate change. The rivers chosen all have significance to the authors, who have knowledge of these rivers through cultural connections as well as through lived experience.

This section of the chapter is written against the backdrop of the unprecedented bushfires that have impacted the Australian continent since September 2019. The 2008 Garnaut Climate Change Review's final report predicted that the effects of climate change on Australia's bushfires would increase over time, stating that the impacts would be directly observable by 2020 (Garnaut). During the bushfire crisis, there has also been a significant acknowledgement of the need to adopt indigenous fire management techniques as a means to mitigate the impact of fire in Australia (Higgins). It is clear that there is much to be learnt from First Nation Australians about the lands and waters of the Australian continent.

The focus is on three rivers:

- *Dhungala* (Murray River), Australia's longest river and part of the Murray-Darling Basin, the driest river system in the world (Murray River);

- *Breimba* (Clarence River) of northern NSW. Apart from the Murray River, it is the *largest* river in mainland Australia south of the Tropic of Capricorn;
- *Molongo* River, which cuts through the middle of Canberra and the Australian Capital Territory (ACT) and feeds into the *Murrumbidgee* River, part of the Murray-Darling Basin.

Each of the rivers discussed have many dreaming stories which explain the flora and fauna that live there, the geological changes and features of the river, and the ecosystems surrounding them. Two of the rivers are connected, as the Murrumbidgee flows into the Murray west of the Alpine high country near Robinvale. The other river, Breimba, or the Clarence, runs from the mountains east to the sea, and over the years there has been talk to diverting the river so it flows into the Murray-Darling.

The Murrumbidgee is part of the Murray-Darling basin, which drains one-seventh of the land mass of the Australian mainland (Pigram, 160–162). The Murray-Darling basin is known as the food bowl of Australia because of the agricultural use of the region. The Murray-Darling Basin also spans four states and the capital jurisdiction – Queensland, New South Wales, Victoria and South Australia. There are over 40 First Nations in the Murray-Darling Basin, each with their own distinct language, culture and stories. Although the river was cared for by many groups, they were all connected through songlines, which told the story of each part of the river, what food was there, what the country looked like and what protocols need to be acknowledged.

We focus in most detail on Dhungala (Murray River) for a discussion about climate change, community, and environmental health and well-being. A key project will be discussed, especially around the context of transdisciplinary collaboration and how to engage appropriately and effectively across cultures and disciplines.

Starting in the Snowy Mountains then creating the border between the eastern states of New South Wales (NSW) and Victoria (Vic), Dhungala crosses many indigenous nations. *Dhungala* (deep waters) is the Yorta Yorta word for the Murray River. The river is known by many names.

Then we track the Breimba (Clarence River) of central New South Wales to further understand how the changes of the rivers over time were mapped and described by the *Bundjalung* people (Hoff). By following the research of geomorphologist Ken McQueen and his work studying the history of the river, a parallel suite of Bundjalung dreaming stories tell of changes to the flow and shape of the river.

The final river, the Molonglo, will also be referenced, in part to address the loss of knowledge and history of the river and the surrounding country. The flooding of the Molonglo river in 1963 to create Lake Burley Griffin in Canberra saw the erasure of limestone caves which held many cave paintings (Jarvis and Dunstan). How can this lost story be repatriated to the story of the city and potential narratives about climate change over time? The waters of the Molongo flow into the Murrumbidgee and then, eventually, the Murray near Robinvale.

Although only a small example of massive cultural and environmental diversity of the Australian mainland, these three examples demonstrate the continuous thread of knowledge about the rivers and the significance of the rivers to ongoing health and well-being of many communities from the people who live in the Murray-Darling Basin and Clarence regions to the communities who rely on the agricultural industry from these areas to provide food.

Yorta Yorta worlds

Each indigenous nation along the river has a responsibility to care for their section of the river and the surrounding land. This knowledge has been passed for tens of thousands of years through song, story and dance.

Yorta Yorta Country includes part of the river country that crosses New South Wales with Victoria, with the Barmah Millewa Forest as an important place for Yorta Yorta. Although an indigenous nation that does not have native title agreements in place, the Yorta Yorta Nation have cared for Dhungala for tens of thousands of years and are still critical stakeholders and cultural owners in the environmental management of the region (State Library of Victoria).

> "We are the river and the river is we."[2]

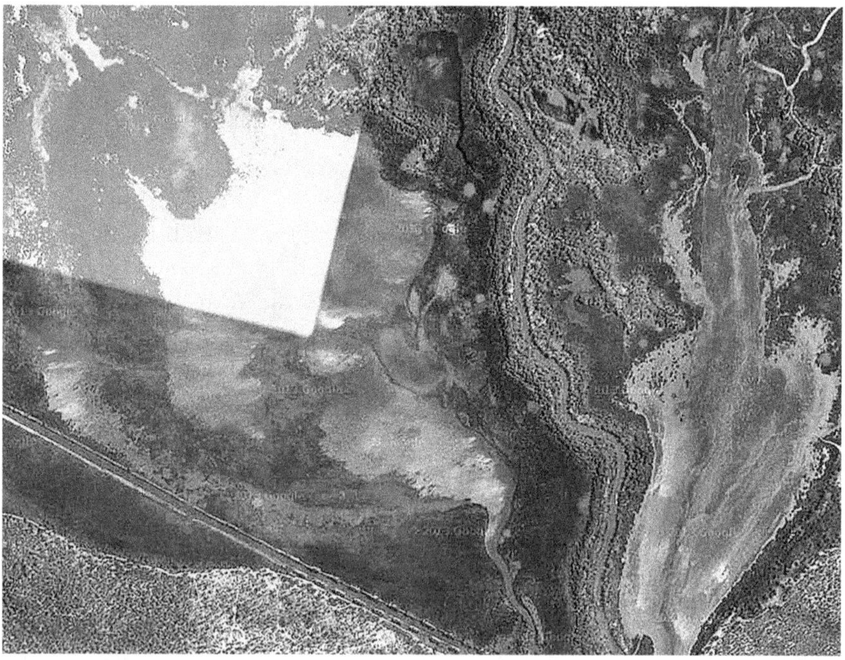

FIGURE 4.1 Lake Barmah. Still image from video "Words for Water: Stage 1."

Source: Tracey M Benson, 2013.

The Yorta Yorta understanding of the river as a metaphor for the body opens up a knowledge base which inherently accepts that biodiversity and ecological balance are critical for not only the environmental health of place, but also of the people who inhabit it. The Yorta Yorta believe that the river is a metaphor of the physical body, referring to Moira and Barmah Lakes as the kidneys that clean the water. Yorta Yorta Elder Uncle Col Walker tells the story in a video online (Walker). As Uncle Col says in the video, this is not only a story of the Yorta Yorta. All along the river system, it is described as different parts of the body – from the "bowels" in the *Coorong* to the intestines by the *Wiradjuri* (Walker). Each part is also part of the songline of the river, shared as a way to understand the river system as the lifeblood for all living things (Kennedy).

Indigenous knowledge of country is increasingly being recognized as having scientific relevance as well as having the potential to guide future decision-making about the care and governance of waterways (Joachim, *et al.*).

There are a number of projects that the Yorta Yorta community have undertaken which strengthen knowledge across generations, ensuring that the community protects their intellectual property and knowledge of country, as well as improving environmental outcomes. A good example is the community mapping project undertaken in 2012–13 intended to support decision-making for the Barmah Millewa Forest (Griggs, *et al.*). Another example is the Youth Journeys that have happened on Country in recent years – an initiative where elders work with young people to pass on culture and knowledge (Goulburn, 61).

For the Yorta Yorta and other communities that represent the geographical area of the Murray-Darling Basin, the connection to the river is clearly tied to cultural well-being. In the Murray-Darling Basin Plan, the role of indigenous communities as partners in the Living Murray Indigenous Partners Program is not clearly defined, though the Barmah Millewa Forest is identified as an iconic site (MDBA, "The Living Murray").

It is critical that the waters can flow so community and environment will flourish. By sharing the deep knowledge of the land and its creatures, the Yorta Yorta and other communities are successfully influencing NRM decision-making at all levels of government. The Murray-Darling Basin Plan is a clear example of the recognition of the important role that indigenous people play.

Breimba and *neyand*

Bundjalung nation is nestled between a number of rivers covering Grafton on the Breimba (Clarence River) in the south, to the Logan River in the north. Breimba winds its way inland as far as the Great Dividing Range, where the river is known as *neyand*. Breimba is the Bundjalung word used to describe the lower reaches of the river near the coast, whereas *neyand* refers to the source of the river, which is located near Tentefield.

This river has the story of Dirrangun, an old woman who wants to keep the water for herself. Artist, designer and Bundjalung woman Bronwyn Bancroft has illustrated the story into a children's book, which has also been used in educational materials (Clever Cookies).

Geomorphologist Ken McQueen has undertaken extensive research on Breimba and noted the significance of the story to the geological history of the river. He states that:

> Many indigenous interpretations of landscapes recognise the importance of rivers. Most invoke carving out of rivers by ancient ('dreamtime') entities. The story of Dirrangun is different, invoking blocking and diversion of drainage in the formation of a river and the related landscape. It indicates a recognition of erosion, incision, deposition and drainage disruption features.
>
> *(30)*

The journey of Dirrangun is also different as a river as it flows to sea on the eastern side of the Great Dividing Range, rather than joining the Murray-Darling Basin to the west.

McQueen also noted that "Indigenous landscape explanations may seem fanciful, but they are often based on insightful geomorphic observations, some even witnessing processes."

Increasingly, there is more Western scientific evidence that can attest to indigenous knowledge of the landscape and land management. Creation stories provide not only a way of understanding the landscape and its specific features, they tie together an understanding of country which is both scientific and spiritual.

Molonglo

The Molonglo river was flooded in 1963 to make way for the creation of Lake Burley Griffin in Canberra, the capital city of Australia. The flooding of the river meant the loss of limestone cave paintings hidden in the banks of the Molonglo. This little-known story of the river does not have a place within the constructs of the national capital and its austere modernity, articulated by the brutalist architecture along the shores of Lake Burley Griffin.

The loss of stories embedded in our landscape is also a loss to our understanding of the ecology of the river and its flows. As McQueen acknowledges, many indigenous stories of the land offer insightful observations about a broad range of areas including erosion, water levels and flood history, abundance of food, changes in the geology and much more. Despite this, and although there has been significant loss to culture and traditional lands, the Ngunawal people continue to demonstrate self-determination through language revival programs and other research that has been compiled about country, in particular, the *Ngunnawal Plant Use* book.

The Australian Capital Territory (ACT) is the regional jurisdiction of Canberra and the surrounding area, and the ACT government recognizes the

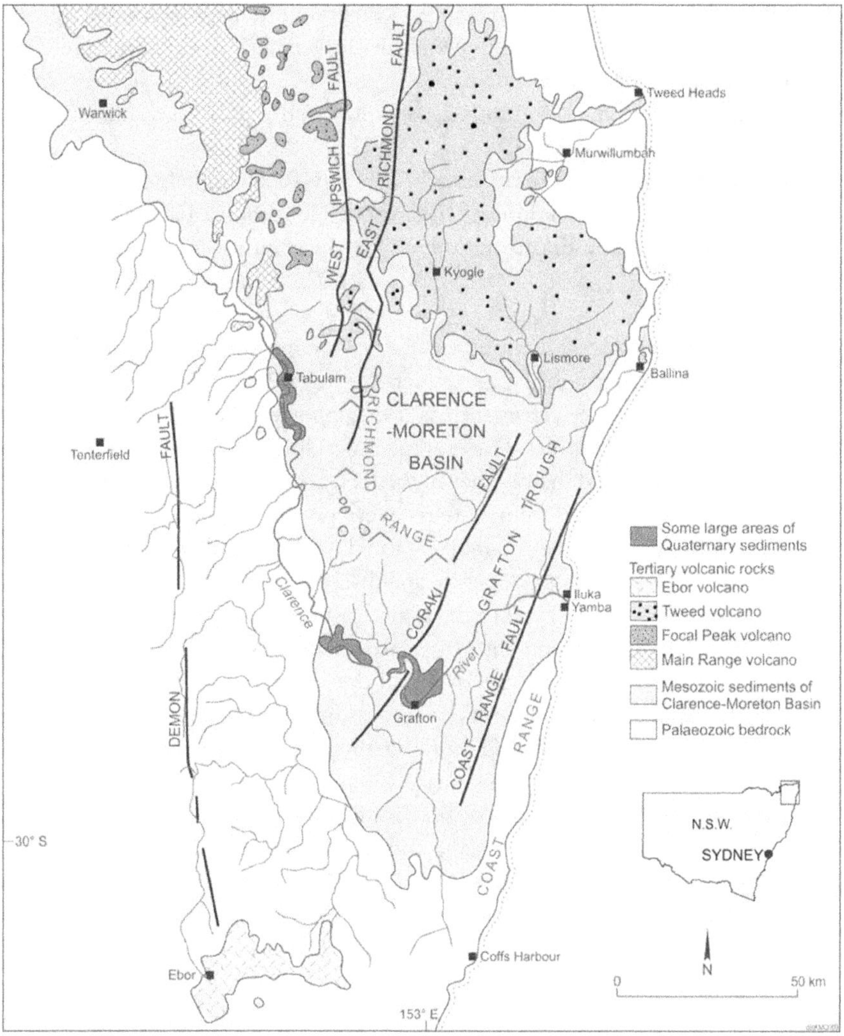

FIGURE 4.2 Map of the Clarence River catchment area showing the main drainage lines and key geological features (from O'Brien et al. 1994). Inset shows area location in northern New South Wales (1:500 000 scale map), Australian Geological Survey Organisation, Canberra.

Source: Adapted by Ken McQueen.

important role that indigenous knowledge and people have in the context of natural resource management (NRM) and education. The ACT government website has a section titled "Indigenous NRM" and states the following:

There is growing recognition that:

- Indigenous people have significant knowledge and skills to offer and responsibilities to fulfill in the protection of Australia's natural resources

- understanding, learning, respecting and applying Indigenous knowledge systems and including Indigenous people in decision-making in natural resource management is key to reconciliation
- strong connection to community, country and culture is fundamental to Indigenous health and well-being
- greater involvement of Indigenous people in NRM will help increase Indigenous employment, improve Indigenous educational achievement; and reducing Indigenous incarceration rates in the criminal justice system.

(ACT Government, Indigenous NRM)

What the website fails to mention is what that contribution gives back to the ACT and broader community, particularly in regards to improved health and well-being. The ACT government has also supported the Ngunawal community through other initiatives. For example, the United Elders Council recently secured funding to build the Ngunawal Bush Healing Farm, which will focus on culturally appropriate well-being and recovery programs (ACT Health).

Much of the ACT is mountainous country and covered in bushland, hence Canberra being known as the "Bush Capital." The abundance of bush also contributes to the diversity of birds in the region, an increasingly rare occurrence in suburban and urban environments. In regards to increasing awareness of the rich diversity of the Murray-Darling Basin and the expertise of traditional owners, more emphasis needs to focus on how that knowledge can help support climate change resilience and the greater value of the river system for all areas of the community.

There is a growing realization that indigenous peoples in Australia were active land managers, shaping the land and waterways to facilitate hunting, fishing and gathering. Bill Gammage's *The Biggest Estate on Earth* does a comparative analysis of the landscape since occupation, noting changes that have occurred since the earliest European recordings of the Australian landscape and vegetation. Although many stories have been lost, one only has to look closely in the landscape to see midden heaps, fish traps and other evidence of a long connection to the land and a story of thriving communities: "the Dreaming is saturated with environmental consciousness" (133).

Compared to Aotearoa New Zealand and the historic legislation granting the Whanganui River legal personhood, Australia still has a long way to go towards respecting our rivers and bushland (Te Aho). Decision-making that improves environmental outcomes needs to be supported by recognizing the significance of waterways and the bush as contributors to the ongoing well-being of the community. The Murray-Darling Basin Plan does attempt to address the importance of the MDB to the health and well-being of the indigenous nations that call the Murray-Darling Basin home, but could have gone further in terms of acknowledging and supporting the deep knowledge held by indigenous peoples. More policies need to reflect the important role that indigenous peoples have in contributing to NRM and climate change policy decision-making. It is encouraging that the Murray Darling Basin Authority recognizes the important relationship that indigenous peoples have to the waterways and also seeks to consider the river system as a whole (MDBA 2016). The next step is educating the community more broadly about how this knowledge can benefit everyone. By studying the history of land and water and the stories attached to them, we can learn more about the places we live and appreciate them more fully.

§

Mātauranga Māori, art and design: unconventional ways for addressing climate change impacts

Huhana Smith, Penny Allan and Martin Bryant

Many coastal *Māori* farming lands, located on lowlands beside the New Zealand coast, are at risk from sea level rise. The Horowhenua region north of Wellington in New Zealand is a prime example. Māori *hapū* (clans) continue to work tracts of ancestral coastal farming land in Horowhenua between the Ōhau and Waikawa Rivers in the southwestern coastal region of *Te Ika ā Maui* (North Island) which, at some point in the next century, will be unarable due to inundation by sea level rise or salinated groundwater tables.

There are problems, too, in the short-term. The viability of efficiency-oriented mono-cultural broad-acre agricultural practices, such as dairying, is questionable: dairy cattle pollute waterways, affecting water resources, depleting *mauri* (vital essence) and devaluing ecosystem services; mono-cultures adversely affect biodiversity; economic constraints encourage financial indebtedness; and broad-acre practices mean that the farms are inaccessible on a day-to-day basis for many of the shareholders, resulting in the marginalization of resource-gathering traditions and land-based customs and values.

Not surprisingly, compelling solutions are complex. Climate change science, geomorphological, archaeo-seismological, hydro-ecological and ecological research provide quantifiable knowledge that is valuable. But the uncertainty of climate change predictions and the complexity of climate change science makes farmers reluctant to prepare and work towards long-term solutions. Perhaps the most pressing demand is to find ways to encourage a cultural shift, persuading conservative Māori farmers and their *whanau* (family) of the opportunities that adaptation can offer for future productivity, environmental health and Māori values.

In this research, we have explored *mātauranga Māori* knowledge to cultivate a more culturally determined approach to land and waterways, and have generated visions, strategies and actions through art and design to provide a more inherently accessible, inclusive and integrative approach to adaptation. This research asks the questions: How can unconventional methods, such as *mātauranga Māori,* complement the scientific method to strengthen the capacity of a place and its people to adapt? How do we bridge the gap between these knowledge systems to make compelling and achievable solutions?

Using a case study farm in the Horowhenua area, our research has applied three Māori methods: *whakapapa* (the interrelated genealogies between all things); *hīkoi* (walking talking hui with knowledge specialists across lands and waterways); and *kōrero tuku iho* (oral narratives of place). Each of these methods is an important form of knowledge, which provides a lens for looking at land/waterways/culture/well-being.

Art and design provide a bridge between these Māori knowledge systems and traditional science. Both focus on local place-specific culture, building synthetic methodologies that enhance the potential for local ownership. And their inherently legible visual language effectively communicates long-term outcomes.

The intent of the research outlined in this section is to encourage adaptation of housing, agriculture and ecosystems to generate beneficial new relationships between settlement form, culture, farming and an enhanced ecological environment. The research method included a series of exhibitions. These showed the potential for engaging Māori farming communities, in the short- and long-term, for regenerating a culturally, environmentally and economically prosperous future.

The case study

This section describes research on the Tahamata Incorporation farm on the Horowhenua coast of *Te Ika ā Maui* (North Island). The farm lies between the polluted waterways of the Kuku Stream, across the banks of the more robust Ōhau River at Te Rauawa, and then stretches south from the Kuku Ōhau estuary and back to the boundaries of the neighbouring *iwi* farm.

Horowhenua is a long coastal plain that once supported one of New Zealand's most significant stretches of wetlands fed by multiple rivers from the Tararua Ranges. In the distant past, these rivers regularly spread rich alluvial soils across the plain and charged extensive wetlands and aquifers, creating an abundant and biodiverse environment for Māori that settled there in the last thousand years. Today, there is still a strong Māori presence on the farmlands, in the marae-centred community settlements, and at *iwi*- (tribal) run *kura* or *wānanga* (Māori schools and university). Farming has been highly productive in the rich soils with plentiful water. But more recently, rivers have been channelized, and water resources have been polluted by fertilizers and concentrations of nutrients. While local *kaitiaki* (guardians) on these farms have reclaimed wetlands and protected estuaries, the remnant natural ecosystems remain fragmented.

Mātauranga Māori

Māori presence imbues spiritual qualities on these lands. Drawn from the creative and artistic intelligence of Ngāti Tūkorehe and Ngāti Wehiwehi, the cultural memory of *kaumātua* (specialist Māori informants) and their oral interpretations, this spiritual knowledge is underpinned by the holistic interrelationships between people, culture and environment. *Mātauranga Māori* is an indigenous worldview that cherishes the connections between everything, and everything we see and do. The separation of the original parents, *Papatūānuku* and *Ranginui*, is a major Māori cosmological event that allowed light to enter the space between their loving embrace. Their separation led to the rise of an extensive range of *atua*, or environmental entities, including those responsible for the growth of forests, insects, animals, fishes and human life. The *Papatūānuku* and *Ranginui* genealogical narrative continues to express the potential of procreation and the promise of new generations to come. As the ultimate parents of *iwi*, *hapū* and *whanau*, they are honoured as the essential expression of *whanaungatanga* that establishes a wide cosmology of peoples, environmental properties, lands, sacred grounds, subsurface waters and waterways.

Many powerful *mātauranga Māori* precepts are present in other Māori cosmological or genealogical narratives. Each narrative is a practical guide for the cultural and philosophical sustenance of Māori, both from a customary perspective and for the benefit of current generations. This research looked at the land through three *mātauranga Māori* knowledge-creation methods.

Whakapapa

Whakapapa means *to layer*. It is the genealogical reference system for hapū and whanau to acknowledge their connections to each other, lands, waterways, ecosystems and areas of spiritual importance. *Whakapapa* suggests that everything is connected and interdependent. While cycles of change occur over time, layers of histories, interactions or activities remain as embodied memories, reinforcing

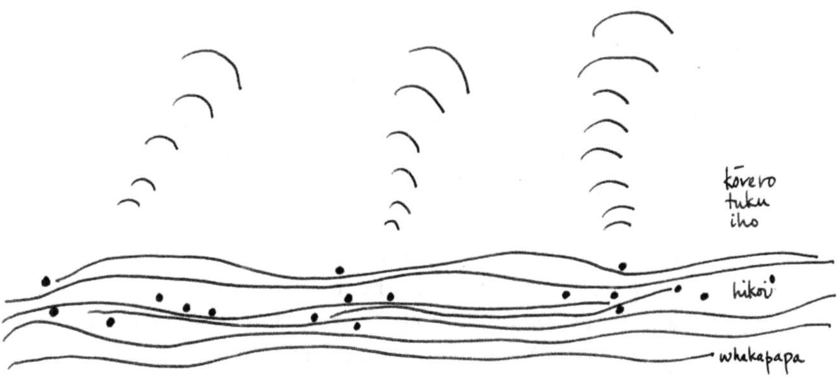

FIGURE 4.3 *Whakapapa* connects land and people. *Hīkoi* is the journey through the land with people. *Kōrero tuku iho* or oral narratives are ways of telling the story of the past for the future.

Source: Image by Martin Bryant (2016).

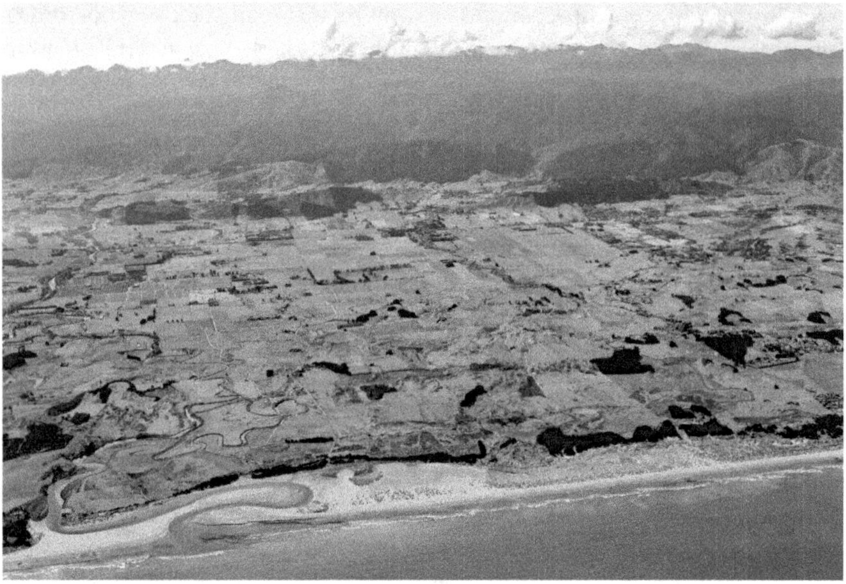

FIGURE 4.4 The large tidal estuary of Kuku Ōhau towards Waikawa, with its changing dunes, rivers and flood plains – the human memories held within represent *whakapapa*.

Source: Aerial photograph by Laurie Cairns, Palmerston North, 28 October 2017.

those connections. The forces of *whakapapa* are embedded in the landscape and the cycles of moons and tides. Māori adapt to these cycles for food and water and well-being.

Hīkoi

Hīkoi is about observing knowledge, sensing it and feeling it, especially through a journey across lands and along waterways. *Hīkoi,* or walking talking *hui* (meetings), is a Māori way of engaging with the land and waterways by being up-close with them, sensing them and drawing upon the wisdom of special places or other moments in the journey across landscapes. Unlike scientific methodologies, the action of walking enables people to link knowledge of place together with experience where lands, waterways, peoples and forces are all interrelated.

Art and design-related research can also be considered as a *hīkoi* process because it entails a continuous and collective unfolding of knowledge that is observed and sensed in ways that are not necessarily linear. Drawing is a way of mapping *hīkoi*, though not as conventional cartography. From up-close and afar, from the past to the future – as if happening at once, these drawings share the experience of the journey: the dunes and the sea, to the rivers and wetlands, and the impact of the wind to the offshore island of Kāpiti and the *pae maunga o Tararua* (Tararua ranges). The drawings reflect the many voices that take part in the journey. They become part of a cognitive map of this place, allowing us to see and sense from many perspectives.

Kōrero Tuku Iho – *oral narratives*

Kōrero tuku iho or oral narratives are *mātauranga Māori* ways of telling stories of the past, the present and the future. They build on whakapapa understandings and expand on the sequences of the *hīkoi*, illuminating moments of journeys through *whenua*. In telling the stories about the past, present and future, oral narratives provide the language and moments that connect things and place with the nature of being. They often highlight key relationships and transformative moments which have acted as thresholds to precipitate new ways of living. For example, when a catch of *tuna* (eels) was landed at full tide to establish a new livelihood, or when a harvest was ruined by flood and *hapū* needed to move cultivations to higher ground.

The exhibition method

The research was framed around four exhibitions spanning 2016 and 2017. The exhibitions were part of an ongoing and iterative process which experimented with ways of communicating the research as it was happening, thereby encouraging communities to play an active role in developing long-term climate change solutions.

The first two, called *Wai o papa*/Waterlands, were held on 31 May and 21 June 2016 at the School of Architecture, Victoria University of Wellington (VUW). The exhibitions were grounded in whakapapa thinking with the community, which established the values and visions that would ensure the farmlands were always abundant and culture thrived despite the uncertainty associated with climate change. There were five visions: bringing *whanau* back to the *whenua*, ensuring healthy land and waterways, improving habitat for *kai moana*, teaching the next generations about their *whakapapa* and keeping the farm economically viable.[3] The visions provided the basis for a series of long-term design strategies to diversify the farm's activities, actively protect all waterways and encourage resettlement on the high ground, thereby enhancing the connection between community, ancestral lands and production without compromising agricultural yield.

The *Wai o papa* exhibitions aimed to disseminate new knowledge on interdependencies between cultural, economic, ecological and climate change issues. The first exhibition drew together data on geological time, geomorphology, groundwater, current climate change science, and Māori understandings of water and tidal cycles. The second focused on the *Maramataka*, a lunar calendar based on oral narratives and observed rituals associated with the rhythms of people, land and water, and the abundance of resources and the phases of the moon. Both exhibitions were foregrounded by the illuminated "Seas will Rise" lightboxes.[4] Located on Vivian Street, a major thoroughfare in Wellington near the exhibition site, the graphics foreshadowed sea level rise through some compellingly simple and direct maps and photographs.

At the same time, inside the School of Architecture, an exhibition of work from fifteen masters students in architecture, landscape architecture and interior architecture showed adaptation strategies for the farm based on an understanding of Māori values developed through *hīkoi*. Students used this knowledge to

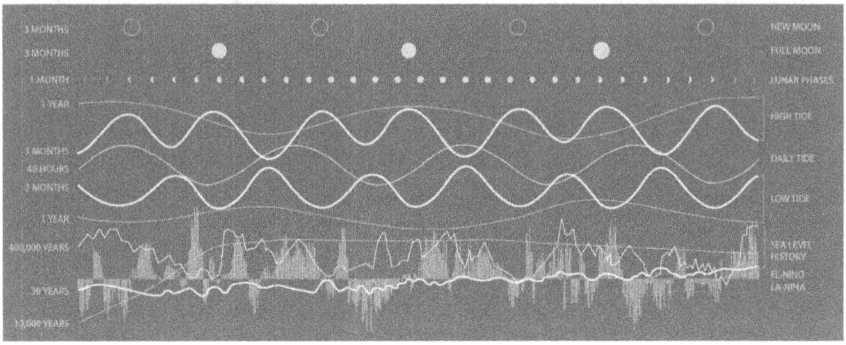

FIGURE 4.5 Detail of consolidated data for the first *Wai o Papa* window exhibition.

Source: Design by Penny Allan and Abdallah Richards 2016.

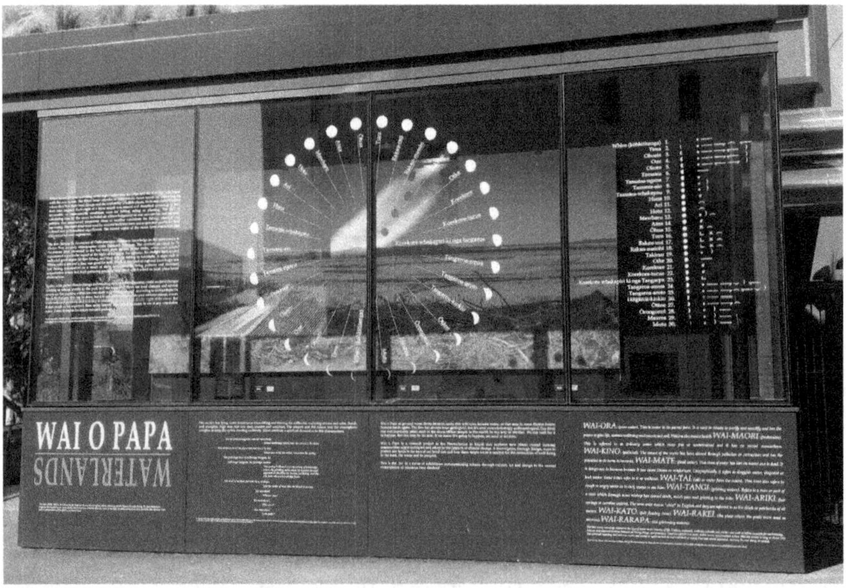

FIGURE 4.6 *Maramataka* exhibition at Te Aro site, Victoria University, Wellington. Design by Huhana Smith and Abdallah Richards.

Source: Photograph by Huhana Smith, 15 September 2016.

envision how climate change might act as a catalyst for positive adaptation. For example: developing *papa-kainga* settlements to re-inhabit new waterfront land on higher ground; recognizing the importance of re-establishing native ecologies; bringing back the birds as well as the *whanau*, *hapū* and *iwi* to their lands; and suggesting how farmland activities might diversify practice and still maintain productivity. Two projects were selected by *iwi* researchers and professional designers to partner with professional firms – Isthmus and Studio Pacific – both based in Wellington, to develop and prepare the work for future exhibitions.

The third and fourth exhibitions events *Whakatairangitia – Rere ki uta, rere ki tai!* Proclaim it to the sea, proclaim it to the land – were held in a collection of disused dairy sheds on the farm and the Dowse Art Museum in Wellington. The dairy sheds exhibition, designed to encourage a full immersion in the site and its whakapapa connections, brought the research together from an indigenous and interdisciplinary perspective. The three sheds by the Kuku Stream were thematically grounded by the concepts of *whakapapa* (interconnected genealogy), *hīkoi* (walking/talking *hui*) and *kōrero tuku iho* (oral narratives of place). The *whakapapa* shed grounded visitors in the site and the vision.

The *hīkoi* shed expressed the journey of the research through an exhibition of art and design and included the students' proposals. It also featured a design proposal to revitalize the *harakeke* industry (NZ flax), planting the farm's waterways and wetland fringes with *harekeke* whilst prioritizing dairying on higher

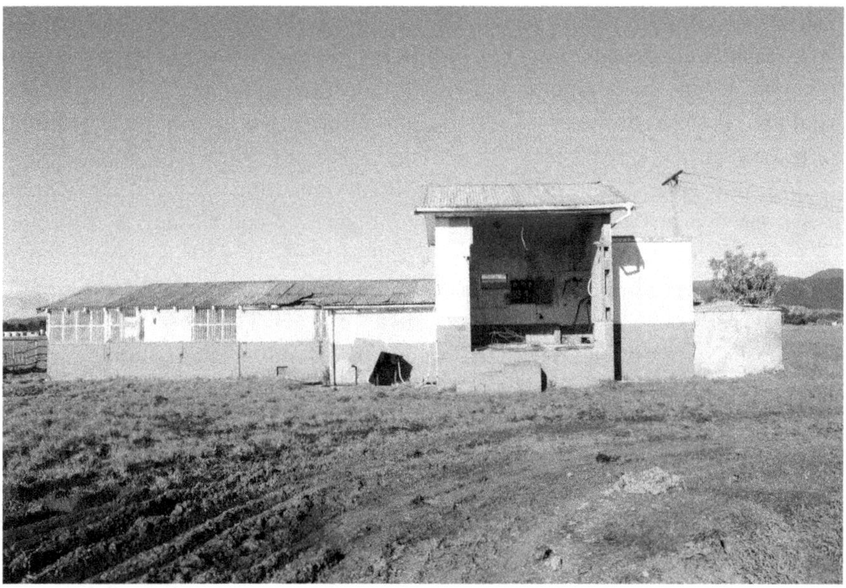

FIGURE 4.7 One of the dairy sheds before clean-up.

Source: Photograph by Huhana Smith, 12 September 2015.

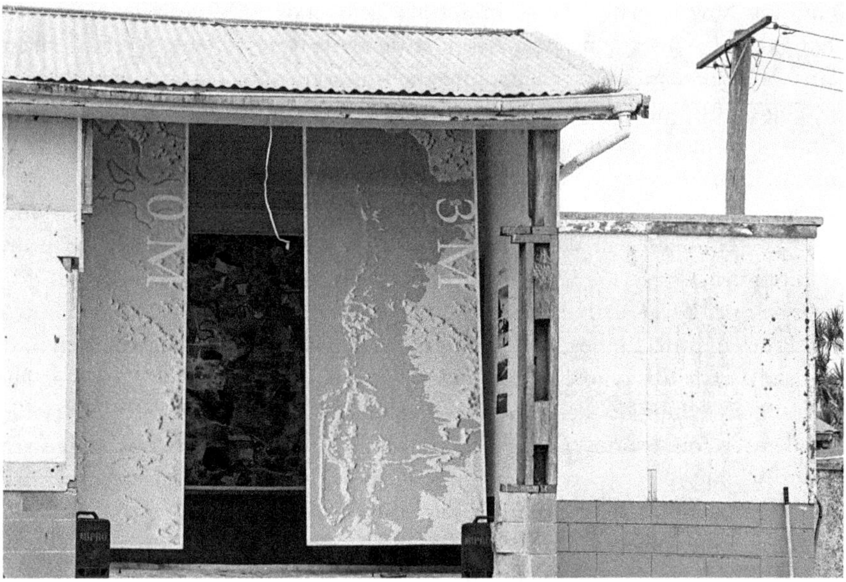

FIGURE 4.8 Close-up of main sea level rise maps in the *Hīkoi*-themed dairy shed.

Source: Photograph by Martin Manning, 11 March 2017.

ground.[5] A revitalized *harakeke* industry works with Māori cultural traditions: in the 19th century, local iwi and hapū harvested it from the margins of wetlands and waterways to develop a significant and sustainable industry around its fibre. Re-establishing *harakeke* also has significant environmental benefits: cleaning polluted waters, encouraging biodiversity, minimizing the effect of erosion and land loss, slowing down flood waters, and mitigating coastal impacts of higher or more salinized water tables.

Art photography by Ann Shelton highlighted the significance of *harakeke* and other native plants as *rongoā* with healing properties used by Māori.

One of the important outcomes of the research was the identification of thresholds[6] – critical ecological signals of the gradual but irreversible shift of the Ōhau River from a river mouth system to an estuary. The erosion and ultimate breach of coastal dunes, an elevated water table, flooding, and failure of vegetation to thrive are thresholds that will herald the shift with wide-ranging impacts on the long-term viability of the farm. The *kōrero tuku iho,* or oral narrative shed, reflected on the nature of these climate change-related thresholds, expressing them visually and asking the question "what would you do if . . .?" to establish a very clear, time-based relationship in climate change terms, between "what might happen" and "what needs to be done."

This linking of thresholds with adaptive responses resonates with the codified knowledge system of the *Maramataka.* However, instead of relating narratives describing past and present relationships between people, land and water, our

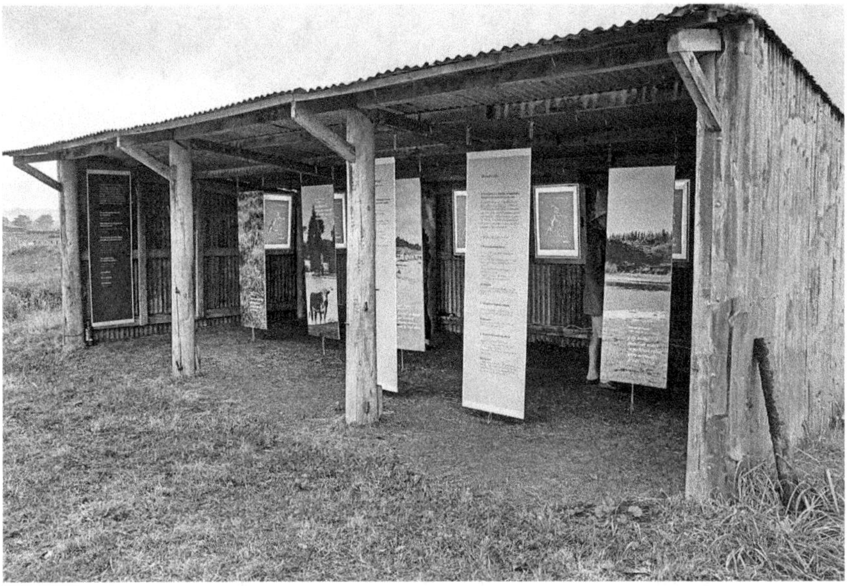

FIGURE 4.9 Thresholds in the *Kōrero Tuku Iho* or oral narratives shed.

Source: Photograph by Martin Manning, 11 March 2017.

intent was to create new narratives, new rituals and new practices, connecting climate change-related phenomena with specific actions and embedding these in cultural practice. We framed the thresholds as catalysts for action and linked them to a toolbox of relatively low-cost adaptation strategies that were tailored to respond to thresholds *as they occurred.* The first strategies related to the protection of arable land from salt water inundation and included revegetation of the coastal dunes and protection of wetlands. The second set of strategies were about accommodating the environmental shift: allowing for flooding in targeted ways, diversifying farming practices and preparing infrastructure for an eventual shift to high ground. A third set of strategies related to resettlement.

All the strategies and tactics were designed to be combined in a variety of ways over time to give farmers multiple options, empowering them to make small incremental changes based on the daily, experiential observations of land and water relationships, rather than having to rely on scientific monitoring or policy-related government interventions.

The fourth exhibition, and the culmination of the research, was part of a group exhibition in 2017 – "This Time of Useful Consciousness: Political Ecology Now" – curated by Senior Curator Melanie Oliver at the Dowse Art Museum in Wellington, featuring a range of artists from Aotearoa New Zealand. While the dairy sheds exhibition took the issues to the local communities, the Dowse exhibition portrayed the intensity of the work in an art gallery, along with other art works.

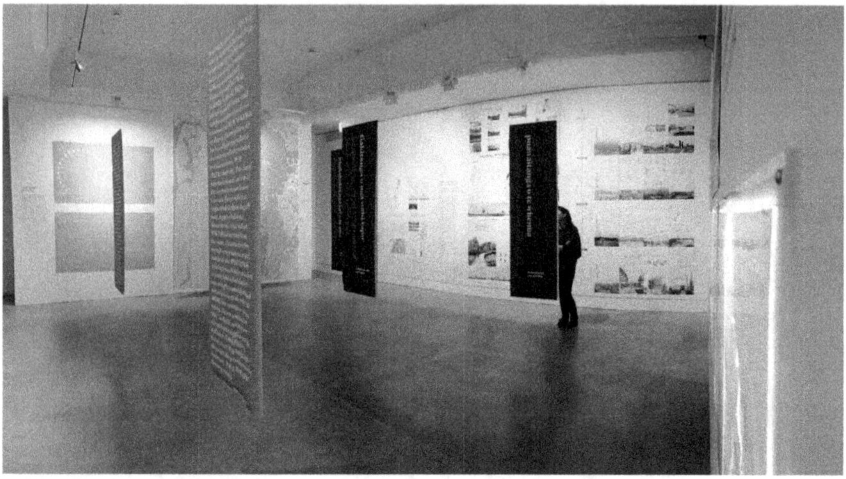

FIGURE 4.10 Views into the exhibition at the Dowse Art Museum, Lower Hutt, Wellington.

Source: Photograph by Huhana Smith, 2017.

Conclusion

The research shows how indigenous knowledge systems, such as *mātauranga Māori*, and cultural values might complement climate change-related scientific research. Because cultural values are close to a community's understanding of land, they provide a way of reinterpreting knowledge derived from Western science to reinforce cultural knowledge. Adaptive actions emerging from this kind of knowledge base are more likely to endure.

The research also shows how qualitative methods like art and design can act concurrently and collectively with scientific research to deliver specific benefits relating to the integration of nature-based and socio-cultural influences, acknowledging risks, vulnerabilities and the adaptive capacity of a place. Exhibitions enhance these methods by grounding knowledge in place and integrating the many threads of scientific and design research to make interdisciplinary knowledge immediately accessible to communities. They act as the basis for transformative change by transferring complex environmental science data and interpretations, enhanced by *mātauranga Māori* and local knowledge of place, to achieve a step change in the public's understanding of environmental science.

Each exhibition speculated on what the past means for the future and how the hapū shareholders or whanau might adapt their activities for the land and waterways in their regions. Such environmental threats are of particular concern to rural Māori, who have customarily had an evolving and adaptable relationship with land, water and resources, but are now very vulnerable to the impacts of climate change and natural hazards on coastal land holding (Allan and Smith, 238). The exhibitions highlighted specific climate change scenarios for the farm, exploring through a range of scales, strategies that defer to the specificity of place and culture. The work exhibited shows that adaptation is going to be slow and incremental, but will need to continue across generations. Importantly, it reinforces the need to engage with climate change adaptation strategies, plan for them and act now.

§

O le ua e afua mai Manu'a: a Samoan indigenous reading of the impacts of climate change[7]

Tamasailau Suaalii-Sauni and Penehuro Fatu Lefale

> *E le o se timu na tō, o le ua e afua mai Manu'a.*
> It is not rain falling, it is a message of love from Manu'a.
>
> *Tupua Tamasese Ta'isi Efi Tui Atua, 2011*

> Those of us from Oceania are already experiencing it [climate change] first-hand. We've seen waves crashing into our homes, and our breadfruit trees wither from the salt and droughts. We look at our children and wonder how they will know themselves or their culture, should we lose our islands.
>
> *Kathy Jetnil-Kijiner, 2014*

> Adaptation to climate change is becoming a Pacific way of life.
>
> *Anne-Marie Tupuola-Plunkett. 2015*

This section explores a Samoan indigenous reading of the impact of climate change on sectors and systems. It reflects on Samoan actions and articulations of environmental responsibility. It examines the relationship between Western science and Samoan indigenous knowledge. It argues that in an indigenous reading, the most powerful motivator to mitigating and adapting to climate change lies in the imperatives of love *(alofa)*: a love for fellow person, for family and for home.

In Samoan indigenous storying, rain is a message of love, a soother of anguish and a sign of providence.[8] Rain (*timu*) are the loving teardrops (*ua*) of Sina, the daughter of the Tui Manu'a.[9] This linking of rain with human teardrops is more than mere poetry, it is evidence of a cultural and moral frame of reference, a way of seeing and being and a way of knowing that draws intimate parallels between the workings of the natural environment and that of the Anthropocene world. It reflects a normative order that privileges the principle of balance: a balance between the sacred or divine and the ordinary or profane; a balance between the feminine and the masculine; between life and death; between taking and giving, remorse and forgiveness (Mead). In a Samoan paradigm, this balance is present in both spiritual and temporal terms, similar to that described by Bernard Narokobi in his thesis on Melanesian jurisprudence (Narokobi). Here, the spiritual and temporal are in constant conversation and interaction with each other. Balance is present when harmony is achieved among all things, animate and inanimate, sentient and insentient. These are core values within what Samoa's head of state philosophically calls, "the Samoan indigenous reference" (Tui Atua 2009b).

In today's world, applying this reference to climate change (*suisuiga o le tau ma le va nimonimo*) conversations seems, for many, foreign and impractical. When read alongside Western scientific definitions of climate change,[10] an unconscious "othering" – or "Orientalism," to use Edward Said's term – occurs. Paradigms of knowing are hierarchialized: those considered most valid in Western colonial terms (Western science) usually sitting at the top and those deemed less valid (indigenous knowledge) are relegated and/or demonized to the bottom (Okere). But the crisis of climate change has now ironically opened up space in global conversations for a more equal, less judgemental dialogue between Western scientific and indigenous paradigms, whereby the value of the latter to humanity is becoming increasingly recognized worldwide, allowing for real conversations beyond the margins between the two (Maradiaga; Roberts).

By definition, a Samoan reading of environmental responsibility privileges a Samoan indigenousness (Alfred) – a reading that brings to the fore Samoan core values, principles and institutions, all of which have remained steadfast over time in their identification of what is uniquely Samoan.

Samoa and Samoan indigeneity

Samoa is a democratic state located in the Pacific Ocean, midway between Aotearoa-New Zealand (NZ) and Hawaii. Her current total human population is just under 200,000. While she is not as at-risk of sinking back into the *vasa* (the sacred space – the Ocean – from which she was born) at the pace suffered by the Marshall Islands, Tuvalu and Kiribati, she is experiencing intense extreme events[11] and concerning levels of soil erosion and coastal retreat. These are linked to post-industrial climatic changes, especially to increases in intensities of natural disasters (e.g., tropical cyclones, tsunamis) and to excessive sand mining (Mimura; Lawson).

Samoa is a member of the United Nations and of the Pacific Islands Forum, and actively participates in global and regional climate change discussions and activities. Her government ratified the UN Framework Convention on Climate Change (UNFCCC) in 1993, and is committed to reducing greenhouse gas emissions, and pursuing low carbon emission developments, especially in the energy sector (Lefale, "AOSIS"; SPREP "PACC"). Samoa recognizes that most of her emissions are from the energy (road transport) and agriculture (livestock farming) sectors, both introduced by modernization and now a normalized – albeit increasingly hazardous – part of Samoan modern lifestyles and culture.

Samoa was the first Pacific Island to gain independence from colonial rule. Her constitution, the supreme law of the land, recognizes Samoan customs and usages.[12] The Village Fono Act gives state recognition to the traditional powers and authority of the village *fono* (council), made up of *matai* (chiefs), who govern

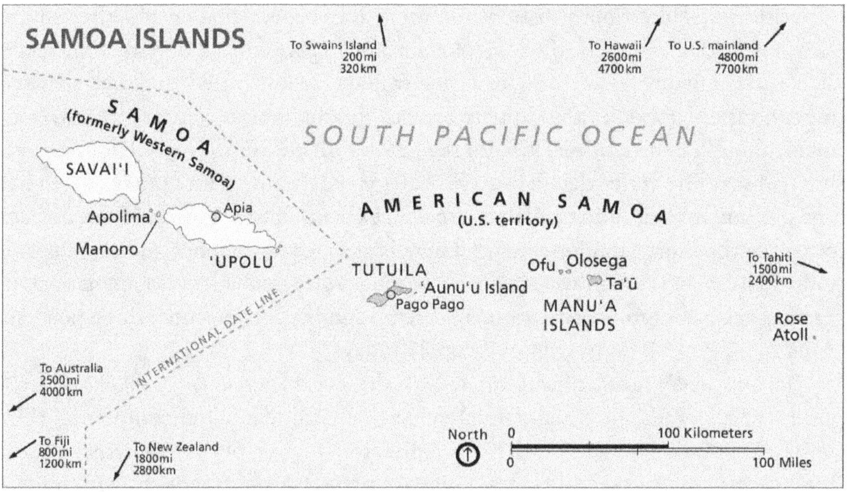

FIGURE 4.11 Samoa and American Samoa.

Source: Map by US National Parks Service.

according to the principles of *faasamoa*, *aganuu* and *agaifanua*, namely Samoan custom law.[13] These three custom law frameworks provide the theoretical basis for a Samoan indigeneity. It is from that base – which today includes a blending of Christian-derived values and beliefs – that contemporary Samoan indigenous environmental conservation practices draw.

A Samoan indigenous paradigm of environmental responsibility

Specifically speaking, within a Samoan indigenous paradigm of environmental responsibility sits six principles: 1. The principle of the *va* – sacred and temporal relational and reciprocal spaces (particularly the *va tapuia* [sacred spaces] and the *va fealoaloa'i* [temporal spaces of reciprocity]) between people and the environment, and between people themselves;[14] 2. The principle of *pule* (i.e., authority bestowed upon one or more to guard and be steward over their family's heritage, including family land and all that pertains to it) (Vaai; Suaalii-Sauni 2010/2011); 3. The principle of *alofa* (love and compassion) (*ibid.*); 4. The principle of *tuā'oi* (of knowing the boundaries of one's rights and responsibilities, inclusive of ideas of *tofi* or divine designations and selfhood) (Tui Atua 2009a; Lefale 2009, 2004); 5. The principles of *tofā mamao* and *faautaga loloto* (i.e., the wisdoms of the long and deep views) (Suaalii-Sauni et al. 2009a); and 6. The principle of *tofā sa'ili* (the Samoan ethic of humility; the view that even though humans will never know absolute truth, they can always be in search of divine wisdom) (*ibid.*). Each of these principles separately and together affirm a philosophy of balance, moderation and harmony for the long-term – for *la longue durée* (Newell).[15]

Within this philosophy, natural resources are believed inalienable (they singularly or together cannot be owned, sold or leased in perpetuity); they are unneglectable (they must always be cared for as one would care for family); and they are to be used to sustain good health. Humans are the guardians or stewards of the environment, alongside their animal kin, for the good of all creation. The environment is a living, breathing entity that shares a genealogy and destiny with humans. Both live and die, and in their living and dying rely on each other to nurture and sustain each other for the short- and long-term. There is an interdependence and an inseparability. This belief system was the basis of Samoa's sustainable development goal, its indigenous paradigm of environmental responsibility, in pre-modern Samoa (Tui Atua 2014, 2009a). So how might it apply today?

Debates over transforming our world: the *2030 Agenda for Sustainable Development of the World*, in which climate change is a fundamental component (UN Sustainable Development Goal 13), has directed intense focus on whether Pacific indigenous environmental philosophies are practical in the modern age (United Nations 2015). For Pacific SIDS, like Samoa, the short answer is yes. Environmental responsibility is as much a state of mind as it is a practice. Indigeneity is not about traditional conservatism, it is about survival and identity – both of which are inextricably linked. Within Pacific island states, indigenous customary ways continue to govern and define everyday life. Many villagers continue to live in traditional *fale* (houses) and the beaches are the economic lifeblood of *aiga* (families: nuclear and extended).

The increased intensities of extreme events and related disasters on the islands has given rise to a re-evaluation of the economic priorities and everyday housing needs of people at national, village and household levels. Adapting lifestyles, attitudes and everyday practices involve literally adapting to a new norm of environmental conditions. As Anne-Marie Tupuola has said, "Adaptation to climate change is becoming a Pacific way of life." This is recognized by the regional Pacific Adaptation to Climate Change (PACC) initiative, administered by the Secretariat of the Pacific Regional Environment Programme (SPREP) and the main climate change donors in the region – the US, NZ, Canada, Japan, China and Australia. Moves to promote a managed return to use of traditional *fale* designs and building materials, and to using trees and natural vegetation to protect sand beaches, are two cases in point.

The resilience of the traditional *fale*

The traditional Samoan *fale*, as depicted in Figure 4.12, is made almost entirely from natural materials, including the sinnet rope (*afa*) that binds the joints of the roof and *pou* (posts). The design was considered by ethnologist Augustin Kramer in the late 1800s as sublime craftsmanship catering superbly to the conditions of Samoa's climate, and to its open extended-family-oriented social living and culture (Kramer). In 2014, Mataafa Autagavaia spoke in support of traditional *fale*

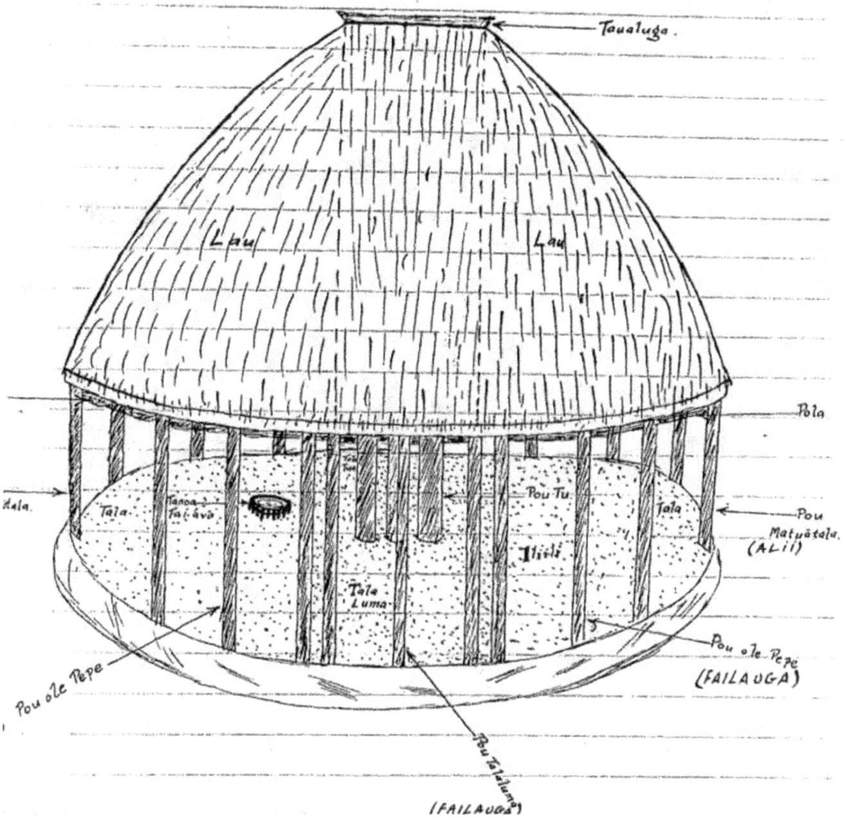

FIGURE 4.12 Traditional Samoan fale with thatched roof and stone foundation.

Source: Image by Te'o Tuvale.

designs, promoting the continued utility value of this design, including its use of natural building materials, arguing that with increased likelihood of cyclones, a thatched roof was less likely to cause serious injury if it were to fly off than a tin roof (Newell 2014, 36–37; UNESCO). Moreover, the open ventilation system of the traditional *fale* design allows for better management of the impact of rising tropical temperature levels, at little to no expense (Wellington Family Centre). It uses less power, lowers costs, reduces reliance on imported petroleum products and lower carbon emissions – a win-win-win, no-regrets policy option measure (IPCC 2001, 474).[16]

A mass return to use of traditional *fale* designs and materials is conceivable only where there is a blend or adaptation of aspects of the *fale palagi* (European house design; Figure 4.13) with that of the traditional *fale*. Such a blend would take into account the climatic, economic and social realities and tastes of Samoa today. In particular, having a kitchen, running water, toilet and bathing facilities

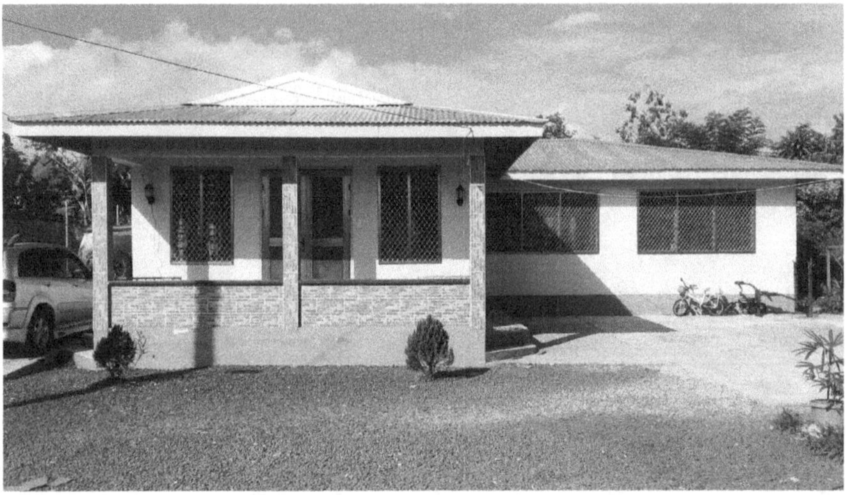

FIGURE 4.13 Example of *fale palagi* or European-designed houses in Samoa, in Vaitele Uta, a non-traditional village in the Faleata district.

Source: Photograph by Cecilia Amosa.

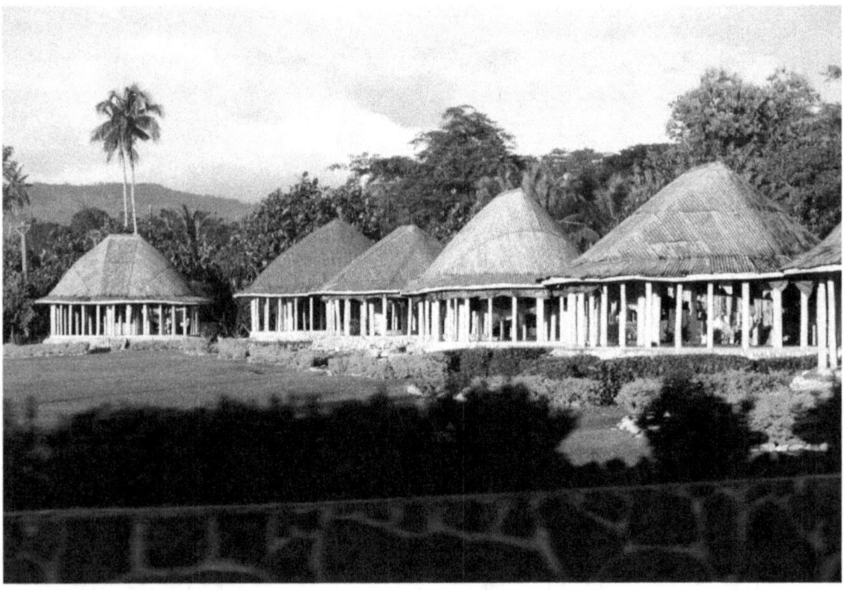

FIGURE 4.14 A cluster of contemporary Samoan *fale*.

Source: Photograph by Faanati Mamea.

in the house, and private individualized sleeping areas, alongside good ventilation, structural durability and aesthetic value. Like the shift to utilize low carbon and renewable energy sources (e.g., solar), while the up-front transition costs to a blended housing design model could be high, the potential long-term economic, environmental and social gains would make the investment worth exploring (IPCC 474). The longevity of the *fale*, like that of the *faasamoa*, is only guaranteed insofar as its core foundations remain relevant.

Protecting beaches protects the ecosystem and family livelihoods

The beach is, like the *fale*, a useful metaphor for human sites of social interaction. However, it is also home to a large number of plants and animals (SPREP; Schupp). For tropical island nations, a healthy beach also provides for a healthy tourism industry, which services the livelihoods of many *aiga* (family). Climate change has had a huge impact on the beach and therefore on family livelihoods. A significant part of climate change adaptation strategies is the state and village councils working together with families, groups and individuals to better achieve awareness and behavioural change at the community ground level.

Basic communication tools, such as good language translation skills (English to Samoan, or formal Samoan to vernacular Samoan) to message climate change actions across the country, coupled with dissemination tools, such as community based workshops and consultations, are fundamental to climate change project success (Figure 4.15) (Samoa Metereology). Projects directly aimed at strengthening beach resilience include reducing levels of sand mining, identifying harmful traditional low-tide fishing methods, planting trees and vegetation in affected areas, monitoring waste disposal, and replacing seawalls with geotextile sand container revetments where possible (Figures 4.16 and 4.17)[17] (Lefale 2004, 2009; Lawson).

Village by-laws have been enacted in Samoa to ensure the state can mobilize appropriate state support for these initiatives (SPREP). The beach that services the popular beach *fale* industry of Manase, in Savaii, has found it difficult to recover quickly from the devastating effects of cyclone Heta over a decade ago. The beach is, according to Taito Vaea Tanumafili, manager of Tanu Beach fales, a family business: "it is our family's and community's livelihood." Over the last 20 years, he has witnessed a steady rise in the sea level. He said that "tropical cyclone Heta transformed the landscape overnight . . . the top layer of the first five metres of the beach were stripped away" (Meyer).

While access to funding resources to facilitate the revetment project for Manase's beach – such as those provided by the USAID Community Adaptation Project (C-CAP) – is important, so too is access to knowledge (traditional and scientific) about basic resource management, conservation and environmental protection acts.

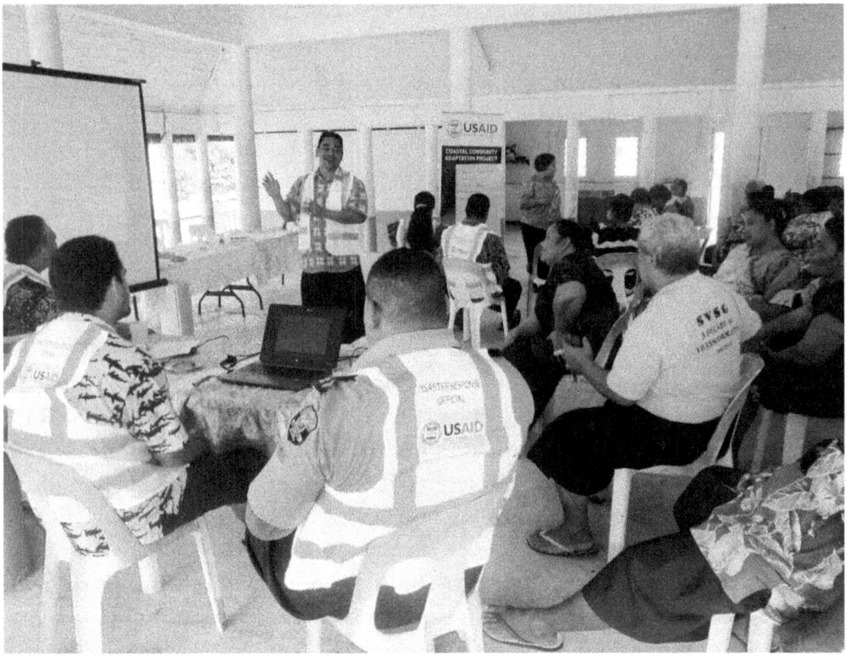

FIGURE 4.15 A weather officer from the Samoan Meteorological Office during his presentation in preparation for the Disaster Simulation Exercise for the Auala village. USAID Coastal Community Adaptation Project, Samoa 2012–17.

Source: Photograph by Cecilia Amosa.

Samoan indigenous principles of resource management, conservation and environmental responsibility are, in theory, central to the *faasamoa* (Samoan culture). In practice, this centrality is not always easily discernible. Not only has climate change opened up global dialogue on indigenous epistemologies, it has, in turn, forced indigenous communities to re-examine and rearticulate for themselves what this means for them in practice, on the ground.

Conclusion – *E le o se timu na tō, o le ua e afua mai Manu'a*

This section sought to explore how Samoan traditional values and principles characterize a Samoan indigenous paradigm of environmental responsibility. It also sought to illustrate how notwithstanding practical challenges, there is a paradigmatic coherence in the *faasamoa* that supports claims of an indigenous Samoan approach. The *fale* and beach examples highlight two of Samoa's approaches to climate change worthy of further research. By continuing to uphold traditional decision-making institutions such as the *aiga* (extended family) and the *faamatai* (chiefly system) in contemporary decision-making settings, Samoa continues to pay more than mere lip-service to its indigenous paradigm.

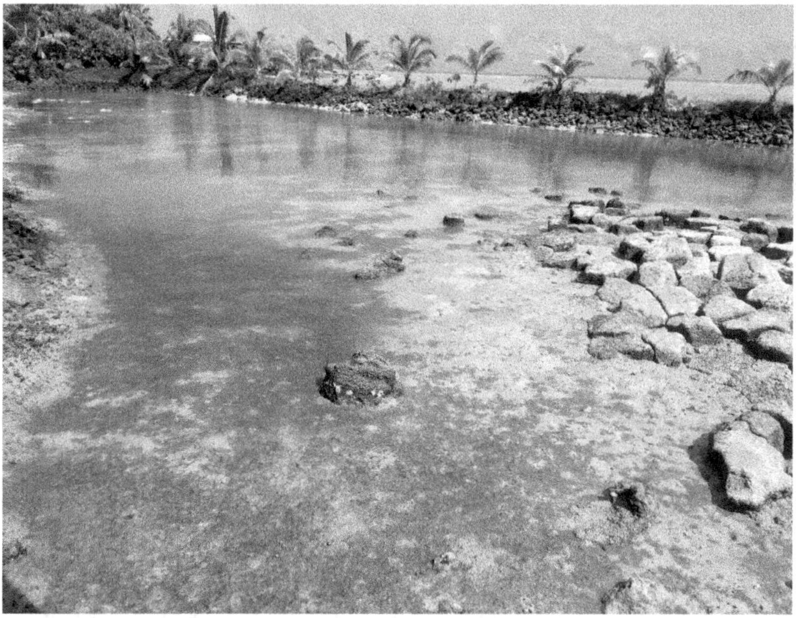

FIGURE 4.16 Seawall, Le Vasa Resort, Mulifanua, Upolu.

Source: Photograph by Penehuro Lefale.

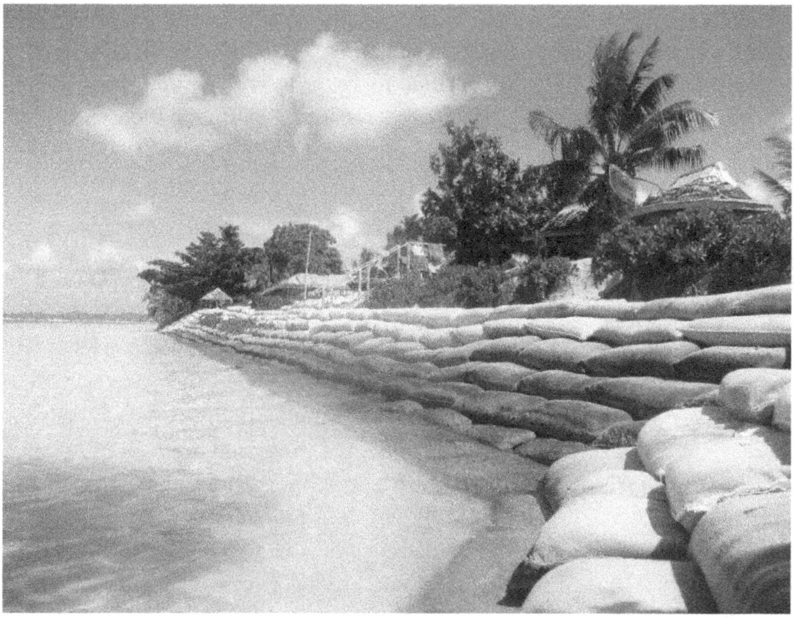

FIGURE 4.17 Geo-textile sand container revetment in Manase, Savaii.

Source: Photograph by Cecilia Amosa.

The plea of Marshallese Kathy Jetnil-Kijiner in 2014 to the leaders of the United Nations is a plea from the heart. It is a plea that is motivated by the same source that caused the tears of Sina to flow from Manu'a to Savaii. It is a plea founded on the reality that taking real action on climate change will happen not just by following scientific formulas for risk assessments, but also by following what motivates the heart – that is, love (*alofa*).

At the water's edge

In examples such as the granting of legal personality for New Zealand's Whanganui River, there are signs neo-colonial and neo-liberal institutions might begin to change. Yet extractive, exploitative imperatives dominate: "the idea of advancing towards a present and future that is but better than the past is not borne out by the statistics" (Anderson, Binney and Harris, 488–9). Indigenous cross-cultural, creative work, storytelling and iterative restoration: all these integral elements, alive in this chapter, must guide leaders, institutions and communities; much of that healing work connects people and environment.

> A Māori [i.e. Indigenous] philosophy of the environment suggests that peaceful, loving and caring ways can be at least as effective as any competitive way of establishing for ourselves meaningful roles in the lands in which we live and also as ways of making those lands better places in which to live for all concerned.
> *(Patterson, 129)*

Sharing and transforming power is central to the adaptive work climate change demands. Indigenous peoples continue to generously share and shape their stories and knowledge as they affirm love, *aroha* and *alofa* as a crucial connective guiding force for all.

Glossary (Māori words)

Hapū:	a collective of *whanau* or family groups
Hīkoi:	to walk
Iwi:	an extended kinship group
Kōrero Tuku Iho:	oral narratives
Kura:	school, education, learning gathering
Marae:	a place where formal greetings and discussions take place
Mātauranga Māori:	the body of knowledge originating from Māori ancestors
Mauri:	life principle, life force, vital essence
Wānanga:	a seminar, conference or forum
Whakapapa:	to recite genealogies, to layer
Whanau:	an extended family group
Whanaungatanga:	a relationship through shared experiences and working together
Whenua:	land

Glossary (Samoan words)

Afa:	sinnet rope
Aiga:	families
Alofa:	love and compassion
Faasamoa, Aganuu and Agaifanua:	components of Samoan custom law
Fale:	Samoan traditional houses
Fono:	council
Matai:	chiefs
Pou:	*fale* posts
Pule:	authority bestowed upon one or more
Timu:	rain
Tofā mamao and faautaga loloto:	the wisdoms of the long and deep views
Tofā sa'ili:	Samoan ethic of humility
Tofi:	divine designations and selfhood
Tuā'oi:	knowing the boundaries of one's rights and responsibilities
Ua:	teardrops
Va fealoaloa'i:	temporal spaces of reciprocity
Va tapuia:	sacred spaces

Notes

1 As recommended by Tyronne Bell – the preferred spelling is Ngunawal, but Ngunnawal is commonly used.

2 The Whanganui Māori saying "I am the river and the river is me" aligns here (Te Aho).

3 These vision statements were forged by Moira Poutama and Aroha Spinks with shareholders at the *wānanga* held at Tukorehe marae, Kuku on 3 November 2016.

4 By Massey University master's student Kevin Cartwright.

5 By VUW graduate student Deborah Scott.

6 Developed with team members Martin Manning, a climate change scientist, and Jane Richardson, a geomorphologist.

7 Given the relatively new area of the interconnection between Western science and the indigenous knowledge field, how to present "results" can make the interpretation and writing exercises difficult. This, and the fact that we are ourselves Samoans, has contributed to our decision to focus this section of the chapter primarily on the Small Island Developing State (SIDS) response of Samoa to climate change. While Samoa is by no means representative of all Pacific SIDS, it does offer some parallels with other Pacific SIDS in terms of both the impact of climate change and its indigenous responses. We are grateful to Espen Ronneberg, Climate Change Advisor at SPREP (Apia office); Mulipola Tainau Ausetalia Titimaea, Assistant Chief Executive Officer of the Meterology Division, Samoa Ministry of Natural Resources and Environment (MNRE); Tile Tofaeono, Principal Scientific Officer, Climate Services, Meteorology Division, MNRE; and Cecilia Amosa, USAID C-CAP Country Mobiliser, for sharing their time and knowledge.

8 The saying *uaga o manū* refers to rain as blessings from the heavens (Tui Atua, 2009b).

9 Mythology tells that the Tui Manu'a was both revered and feared for his supernatural powers. Tui Manu'a is the title of the highest-ranking chief of the Manu'a islands. The Manu'a islands are part of current day American Samoa. In ancient times, the Tui Manu'a was part of the Tui regime which included the Tui Atua and Tui Aana (of

Samoa), the Tui Fiti (of Fiji), the Tui Toga (of Tonga) and the Tui Uea (of Wallis and Futuna). Briefly, the love story goes that Sina had fallen in love with a young man from Savaii, named Lemanunu, and had run off with him to Savaii. Unfortunately, Lemanunu was not deemed worthy enough to be her suitor. Sina knew that once her father found out what she had done he would be enraged. Sure enough, upon finding out, her father ordered her sister, Aolele, to go to Savaii to bring her back. When Aolele and her entourage arrived at Lemanunu's village, they found the two very much in love and full of anguish. But everyone knew that Sina had to obey her father's wishes, especially if she did not want any harm was to come to Lemanunu and his family. Her parting words were: "When you see the rain coming from Manu'a, it is not rain, they are tears (*ua*) of my undying love." Sina sacrificed her happiness for Lemanunu and his family's safety. It was once not uncommon to hear the elders in Savaii recite this saying whenever they saw rain coming from the east. (The content offered for this version of the story was drawn from Tui Atua, 2009b.)

10 Climate change refers to "a change in the state of the climate that can be identified (e.g. using statistical tests) by changes in the mean and/or the variability of its properties, and that persists for an extended period, typically decades or longer" (UN Intergovernmental Panel on Climate Change (IPCC) definition). This usage differs from that in the United Nations Framework Convention on Climate Change (UNFCCC), where climate change refers to "a change of climate that is attributed directly or indirectly to human activity that alters the composition of the global atmosphere and that is in addition to natural climate variability observed over comparable time period." (UNFCC)

11 A weather or climate event that is rare at a particular place (and, sometimes, time of year) including, for example, heat waves, cold waves, heavy rains, periods of drought and flooding, and severe storms. Definitions of rare vary, but an extreme weather event would normally be as rare as or rarer than a particular percentile (e.g., 1st, 5th, 10th, 90th, 95th, 99th) of a probability density function estimated from observations expressed as departures from daily or monthly means (National Academies of Sciences).

12 1960 Constitution of the Independent State of Western Samoa. The preamble states: "Whereas sovereignty over the Universe belongs to the Omnipresent God alone, and the authority to be exercised by the people of Western Samoa within the limits prescribed by His commandments is a sacred heritage . . . the Leaders of Western Samoa have declared that Western Samoa should be an Independent State based on Christian principles and Samoan custom and tradition." In 1997, the name "Western Samoa" was changed to "Samoa" by the 1997 Constitution Amendment Act (No. 2) No. 15 (So'o).

13 *Faasamoa* literally means "to be of (*faa*) *Samoa*"; *aganuu* literally means the customs or norms (*aga*) of a village (*nuu*); and *agaifanua* literally means the customs or norms (*aga*) pertaining to (*i*) a specific land area (*fanua*) (See Vaai, 1999; Suaalii-Sauni, 2010/2011, 2017).

14 There is a growing literature on this principle (Suaalii-Sauni 2017b).

15 In the sense of *longue durée* used by Jennifer Newell. Newell draws on the work of Fernand Braudel. The *World Heritage Encyclopaedia* helpfully notes that the concept of "longue durée" was developed by French historian Fernand Braudel in his examination of social scientific methodologies. It notes that "[I]n the longue durée of economic history," for example, "beyond, or beneath, the cycles and structural crises, lie 'old attitudes of thought and action, resistant frameworks dying hard, at times against all logic.'" Retrieved June 2017, online at: www.worldebooklibrary.org/articles/eng/longue_dur%C3%A9e.

16 No-regrets options are by definition GHG emissions-reduction options that have negative net costs because they generate direct or indirect benefits that are large enough to offset the costs of implementing the options.

17 Or with elcorock seawalls. After an extreme tide event which displaced some of the geotextile sand containers on the eastern end of the revetment mid-2016, the USAID C-CAP project installed an elcorock seawall (personal email communication from Cecilia Amosa, July 2017).

Bibliography

'Aboriginal Victoria'. Visit Victoria. www.visitmelbourne.com/Regions/The-Murray/Things-to-do/Aboriginal-Victoria (Accessed 2 July 2017).

ACT Government. 'Indigenous Natural Resource Management.' www.environment.act.gov.au/cpr/indigenous_nrm (Accessed 23 June 2017).

ACT Health. ACT Government. 'Ngunnawal Bush Healing Farm.' www.health.act.gov.au/public-information/consumers/health-infrastructure-program/current-projects/ngunnawal-bush-healing (Accessed 22 June 2017).

Alfred, Taiaiake and Jeff Corntassel. 2005. 'Being Indigenous: Resurgences against Contemporary Colonialism." *Government and Opposition: An International Journal of Comparative Politics*, 40: 4, 597–614.

Allan, Penny and Huhana Smith. 2017. 'He Whakawhiti Kōrero: A Conversation about a Collaboration in Art, Design and the Environment.' In C. McCarthy and M. Stocker (eds.) *Colonial Gothic to Māori Renaissance: Essays in Memory of Jonathan Mane-Wheoki*. Wellington: Victoria University Press, 239.

Anderson, Atholl, Judith Binney and Aroha Harris. 2014. *Tangata Whenua: An Illustrated History*. Wellington: BWB.

Clever Cookies. 2009. 'Bronwyn Bancroft's Aboriginal Art.' https://clevercookies.edublogs.org/2009/03/20/bronwyn-bancrofts-aboriginal-art/ (Accessed 23 June 2017).

De Brouwer, Gordon. 2017, 'Effective Frameworks for Natural Resource Management.' presented at the Australian Agricultural and Resource Economics Society Conference, February 8. www.environment.gov.au/about-us/departmental-structure/secretary/effective-frameworks-natural-resource-management (Accessed June 21 2017).

Gammage, Bill. 2011. *The Biggest Estate on Earth: How Aborigines Made Australia*. Sydney: Allen and Unwin.

Garnaut, Ross. 2008. *Garnaut Climate Change Review, Final Report*. Cambridge: Cambridge University Press.

Goulburn Broken Catchment Management Authority. 2015. *Annual Report 2014–15*. 'Yorta Yorta Youth Journey a Huge Success.' www.gbcma.vic.gov.au (Accessed 23 June 2017).

Griggs, David, Amanda Lynch, Lee Joachim, Xuan Zhu, Carolina Adler, Zac Bischoff-Mattson, Pan Wang and Tahl Kestin. 2014. *Learning from Indigenous Knowledge for Improved Natural Resource Management in the Barmah Millewa*. Final Report to the Victorian Centre for Climate Change Adaptation Research. www.vcccar.org.au/publication/final-report/learning-from-indigenous-knowledge-for-improved-natural-resource-management (Accessed 3 July 2017).

Higgins, Isabella. 2020. 'Indigenous Fire Practices Have Been Used to Quell Bushfires for Thousands of Years, Experts Say.' *ABC News*. www.abc.net.au/news/2020-01-09/indigenous-cultural-fire-burning-method-has-benefits-experts-say/11853096 (Accessed 11 January 2020).

IPCC, *Mitigation*. 2001. *Working Group III*. Accra: Ghana. www.ipcc.ch/ipccreports/tar/wg3/index.php?idp=292 (Accessed 6 July 2017).

Jarvis, Amy and Jack Dunstan. 2016. 'Before ANU Was Here.' *ANU Reporter*. http://reporter.anu.edu.au/anu-was-here%E2%80%A6 (Accessed 2 July 2017).

Jetnil-Kijiner, Kathy. 2014. *UN Climate Leaders' Summit*. www.youtube.com/watch?v=L4fdxXo4tnY (Accessed 4 July 2017).

Joachim, Lee, Amanda H. Lynch, David Griggs and Jackie Walker. 2013. 'The Role of the Yorta Yorta People in Clarifying the Common Interest in Sustainable Management of the Murray: Darling Basin, Australia.' *Policy Sciences*, 46: 2, 109–123. doi:10.1007/s11077-012-9164-8 (Accessed 3 July 2017).

Kennedy, Brendan. 2017. *Conversation with Tracey Benson*. Renmark, South Australia, 29 April.

Kramer, Augustin. 1995. 'House Building.' In T. Verhaaren (Tr.) *The Samoa Islands: An Outline of a Monograph with Particular Consideration of German Samoa*, Volume 2. Auckland: Polynesian Press, 259–280.

Lawson, Sawyer. 2011. 'Seawalls in Samoa: A Look at Their Environmental, Social and Economic Implications.' *SIT Independent Study Project (ISP) Collection*. Paper 1058. http://digitalcollections.sit.edu/isp_collections/1058.

Lefale, Penehuro Fatu. forthcoming. "AOSIS: Where to from Paris? Climate Clubs and the Alliance of Small Island States (AOSIS).' In H.E. Ott, et al. (eds.) *Climate Alliances Après Paris: The Potential of Pioneer Climate Alliances to Contribute to Stronger Mitigation and Transformation*. Bonn: German Development Institute, 11–22, www.die-gdi.de/en/.

Lefale, Penehuro Fatu. 2004. 'A Samoan perspective on the role of traditional knowledge in advancing our understanding of climate change'. In E. Helander and T Mustonen. (eds.). *Snowscapes, dreamscapes. snowchange: Book on community voices of change*. Ser C., Fram Oy, Vaasa, Finland: Tampere Polytechnic Publications.

Lefale, Penehuro Fatu. 2009. 'An epic quest for the Samoan indigenous reference'. In: *Su'esu'e Manogi: In search of fragrance: Tui Atua Tupua Tamasese Ta'isi Efi and the Samoan indigenous reference*. (Suaalii-Sauni, T. et a., eds). Apia: National University of Samoa.

Maradiaga, Oscar Andrès Cardinal Rodriguez. 2014. 'Sustainable Humanity, Sustainable Nature.' Paper for the Sustainable Humanity, Sustainable Nature: Our responsibility symposium, Pontifical Academy of Social Sciences, Extra Series 41, Acta 19, Vatican City. www.pas.va/content/dam/accademia/pdf/es41/es41-rodriguezmaradiaga.pdf (Accessed 2 July 2017).

McQueen, Ken G. 2017. 'Formation of the Clarence River in SE Australia and the Story of Dirrangun: Ethnogeomorphology and Relevance to Landscape Evolution and Climate Change.' 17th Australia New Zealand Geomorphology Group Conference, Greytown, N.Z. www.anzgg.org/images/ANZGG_Program_and_Abstracts_low_res.pdf (Accessed 3 July 2017).

Mead, A.T.-P. 1998. '"Sacred Balance': Global Biodiversity Assessment: Cultural and Spiritual Values of Biodiversity', United Nations Environment Programme (UNEP). www.yumpu.com/en/document/view/11402568/pdf-sacred-balance-aroha-te-pareake-mead-final/5 (Accessed June 2017).

Meteorology Services and Ministry of Natural Resources and Environment. 2017 (circa). *Samoa Meteorology Glossary in Samoan and English*. Apia: Samoa Meteorology Service & MNRE.

Meyer, L. 2015. 'Saving Pacific Island Countries, One Community at a Time.' Frontline (Online Edition). www.usaid.gov/news-information/frontlines/climate-change-2015/saving-pacific-island-countries-one-community-time?page=1 (Accessed 5 July 2017).

Mimura, Nobuo. 1999. 'Vulnerability of Island Countries in the South Pacific to Sea Level Rise and Climate Change.' *Climate Research*, 12, 137–143.

Ministry of Education, Sports and Culture. Matagaluega o Autalavou Taaloga ma Aganuu. 1999. 'Traditional Samoan Fale.' Apia: Samoa.

Murray-Darling Basin Authority (MDBA). n.d. 'The Living Murray Program.' www.mdba.gov.au/managing-water/Environmental-water/delivering-environmental-water/living-murray-program. (Accessed 21 June 2017).

Murray-Darling Basin Authority (MDBA). 2016. *Our Water, Our Life: An Aboriginal Study in the Northern Basin*. Canberra: MDBA. www.mdba.gov.au (Accessed 23 June 2017).

Murray Regional Tourism Board. 'The Murray.' www.visitthemurray.com.au/about-the-region/fast-facts (Accessed 3 July 2017).

Narokobi, Bernard. 1989. *Lo bilong yumi yet: law and custom in Melanesia*. Suva, Fiji: The Melanesian Institute for Pastoral and Socio-Economic Service and the University of the South Pacific.

National Academies of Sciences, Engineering, and Medicine. 2016. *Attribution of Extreme Weather Events in the Context of Climate Change*. Washington, DC: The National Academies; Press. doi: 10.17226/21852 (Accessed 7 July 2017).

National Park Maps. 'Regional Samoan Map.' www.nps.gov/carto/hfc/carto/media/NPSAmap2.jpg (Accessed 8 December 2019).

O'Brien, P.E., Korsch, R.J., Wells, A.T., Sexton, M.J. and Wake-Dyster, K. 1994. 'Structure and Tectonics of the Clarence-Moreton Basin.' In A.T. Wells and P.E. O'Brien (eds.) *Geology and Petroleum Potential of the Clarence-Moreton Basin, New South Wales and Queensland*. Bulletin 241. Canberra: Australian Geological Survey Organisation, 195–217.

Okere, Theophilus. 2005. 'Is There One Science, Western Science?' *Africa Development*, 30: 3, 20–34.

Patterson, John. 2001. *People of the Land: A Pacific Philosophy*. Palmerston North: Dunmore Press.

Pigram, John J. 2007. *Australia's Water Resources: From Use to Management*. Collingwood, Victoria: CSIRO Publishing.

Referendum Council's Indigenous Steering Committee. 2017. 'Aboriginal and Torres Strait Islanders Peoples from Across Australia Make Historic Statement.' Press release. 26 May. www.referendumcouncil.org.au (Accessed 23 June 2017).

Roberts, Mere. 1998. 'Indigenous Knowledge and Western Science: Perspectives from the Pacific.' In *Science and Technology Education and Ethnicity: An Aotearoa/New Zealand Perspective*. Wellington: Royal Society of New Zealand, 59–75.

Samoa Meteorology Services and Ministry of Natural Resources and Environment. c.2017. *Samoa Meteorology Glossary in Samoan and English*. Apia: SMS & MNRE.

Schupp, C. 2013. *Coastal Management Adaptation to Climate Change using Ecosystem-Based Adaptation Strategies (Samoa Component)*. Apia: Ministry of Natural Resources and Environment. www.mnre.gov.ws/wp-content/uploads/2016/11/Coastal-Management-Plans-for-Manase-20130821-with-figures.pdf (Accessed on June 2017).

So'o, A. 2008. *Democracy & Custom in Sāmoa: An Uneasy Alliance*. Suva, Fiji: IPS Publications, University of South Pacific.

SPREP. (n.d.). *Pacific Adaptation to Climate Change, Samoa Report of in-Country Consultations*. Apia: SPREP. www.sprep.org/att/publication/000663_Samoa__National_PACCReport_Final.pdf (Accessed 23 June 2017).

State Library of Victoria. 'Native Title & the Yorta Yorta.' n.d. http://ergo.slv.vic.gov.au/explore-history/fight-rights/indigenous-rights/native-title-yorta-yorta.

Suaalii-Sauni, Tamasailau. 2010/2011. '"It's in Your Bones!": Samoan Custom and Discourses of Certainty.' In R. Benton (eds.) *Yearbook of New Zealand Jurisprudence*, 13&14, 70–88. Hamilton: University of Waikato.

Suaalii-Sauni, Tamasailau. 2017a. 'Legal Pluralism and Politics in Samoa: The Faamatai, Monotaga and the Samoa Electoral Act 1963.' In *Small States in a Legal World*. Switzerland: Springer, 165–190.

Suaalii-Sauni, Tamasailau. 2017b. 'The va and Kaupapa Māori Research.' In T.K. Hoskins and A. Jones (eds.) *Critical Conversations in Kaupapa Māori*. Wellington: Huia Publishers, 161–178.

Te Aho, Linda. 2014. 'Ruruku Whakatupua Te Mana o te Awa Tupua – Upholding the Mana of the Whanganui River.' *Māori Law Review*: 5. http://maorilawreview.co.nz (Accessed 24 June 2017).

Tui Atua, Tupua Tamasese Ta'isi Efi. 2009a. 'Climate Change and the Perspective of the Fish.' Keynote address for Stars of Oceania Summit, University of Hawaii, Oahu, Hawaii. www.head-of-state-samoa.ws/speeches_pdf/Climate%20Change_April%20 2009.pdf (Accessed 5 July 2017).

Tui Atua, Tupua Tamasese Ta'isi Efi. 2009b. 'Water and the Samoan Indigenous Reference.' In T. Suaalii-Sauni et al. (*eds.) *Su'esu'e Manogi: In search of fragrance: Tui Atua Tupua Tamasese Ta'isi Efi and the Samoan indigenous reference.* Apia: National University of Samoa. Chapter 17.

Tui Atua, Tupua Tamasese Ta'isi Efi. 2011. '*E le o se timu na to, o le ua na afua mai Manu'a*: A message of love from *fanauga*'. Keynote address. Professor James Ritchie Memorial Lecture Series, University of Waikato. Hamilton (NZ), 23 February. www.head-of-state-samoa.ws/speeches_pdf/Ritchie%20Lecture%20FINAL4%20with%20 dedication%2006%2004%2011.pdf (Accessed June 23 2017).

Tui Atua, Tupua Tamasese Ta'isi Efi. 2014. 'Disaster Preparedness and "the Indigenous" in the Pacific.' Opening remarks for Parallel Event, Pacific Research and Policy Centre and the Joint Centre for Disaster Research, Massey University. UN Conference on Small Island Developing States, Apia, Samoa. www.headofstate.ws/disaster-preparedness-and-the-indigenous-in-the-pacific/ (Accessed 23 June 2017).

Tupuola-Plunkett, Anne-Marie. 2015. 'Resilience Despite Adversity: Adapting Indigenous Communities to Climate Change in the Pacific Region.' *Minority Voices Newsroom.* www.minorityvoices.org/news.php/fr/1809/swm-2015-resilience-despite-adversity-adapting-indigenous-communities-to-climate-change-in-the-pacif (Accessed 23 June 2017).

Tuvale, Te'o. 1918. 'An Account of Samoan History Up to 1918.' http://nzetc.victoria.ac.nz/tm/scholarly/tei-TuvAcco.html. CC. by 3.0 (Accessed 12 July 2019).

UNESCO. 2013. *Traditional Knowledge for Adapting to Climate Change: Safeguarding Intangible Cultural Heritage in the Pacific.* Apia/Daejeon: UNESCO & ICHCAP.

UNFCC. 2011. 'Fact Sheet: Climate Change Science: The Status of Climate Change Science Today.' Geneva: United Nations Framework Convention on Climate Change. https://unfccc.int/files/press/backgrounders/application/pdf/press_factsh_science.pdf

UNIPCC. 2014. *Summary for Policymakers.* Field, et al (eds) *Climate Change 2014: Impacts, Adaptation, and Vulnerability.* Cambridge, United Kingdom and NewYork, USA: Cambridge University Press. www.ipcc.ch/site/assets/uploads/2018/02/ar5_wgII_spm_en.pdf

United Nations. 2015. *Transforming Our World: The 2030 Agenda for Sustainable Development.* https://sustainabledevelopment.un.org/post2015/transformingourworld (Accessed 7 July 2017).

Vaai, S. 1999. *Samoa Faamatai and the Rule of Law.* Apia: National University of Samoa.

Walker, Col. 'The Kidney Story.' *Online video.* http://debraneys.com/stories-of-the-yorta-yorta-people/the-kidney-story.html (accessed 2 July 2017).

Wellington Family Centre & Afeafeovaetofaga i Vaialua (Pacific Academy of Cultural Restoration, Research and Development). 2009. 'Indigenous Housing as a Solution to Climate Risks' Project." www.slideshare.net/Afeafeovaetoefaga/indigenous-housing-as-a-solution-to-climate-risks (Accessed 20 June 2017).

5

INDIGENOUS RELIGIONS OF OCEANIA

Australia, New Zealand and the South Pacific Islands

James L. Cox

The belief systems and social structures within indigenous societies revolve around ancestors, and hence, the religions of such societies are localized as opposed to universal. Indigenous societies also exclusively relate to a particular geographical place fixed in the tradition by their quasi-legendary myths. Unlike religions with universal cosmologies, one can never be converted into an indigenous religion, since ancestors only belong to precise kinship groups that operate within restricted and delineated spatial locations (Platvoet 1992: 21–22; Cox 2007: 61–71).

In this chapter, I apply this limited definition of indigenous religions to the regions of the South Pacific, often referred to as Oceania, by focusing on the Aboriginal religions of Australia, the Māori of Aotearoa (New Zealand) and the indigenous religions of the southern Pacific stretching from Melanesia to the eastern Polynesian Islands. Of course, wide variations can be found among the societies within this vast area. In their ground-breaking book, *The Religions of Oceania*, Tony Swain and Garry Trompf (1995: 1) admit that the "south-west Pacific has hundreds of language and cultural groups, not to mention an immense history which will mostly remain forever concealed." Australian Aboriginal societies are distinct from those in New Zealand and the South Pacific islands, but even within the Australian continent, indigenous societies and languages are numerous and diverse. Despite the wide variations within particular contexts across the whole of Oceania, I demonstrate in this chapter that it is possible to make limited generalizations based on the primary characteristics of indigenous religions: kinship and locality.

I begin by examining the Australian case before moving to New Zealand, followed by the southern Pacific Island region. I will identify issues in the study of the religions of each area, followed by a description of some of the key themes emerging out of selected myths, beliefs and rituals in the particular regions under study. I conclude the chapter by analyzing the significance of the specific religions I have described for understanding indigenous religions generally, while at the same time drawing attention to their dynamic, adaptive and adoptive nature.

Australia: background

Before the arrival of European settlers, the indigenous population of Australia was estimated to have numbered around 300,000, although some scholars suggest that the number may have been closer to 750,000 (Charlesworth 2005: 2). There were as many as 500 different groups or sub-groups covering the 7,700,000 square kilometres which make up the land mass of the Australian continent. Each group had its own language or dialect, and lifestyles and cultural traditions that differed according to the region in which they lived. The 2011 Australian national census reported that of the just over 25 million inhabitants in today's Australia, around 550,000 consider themselves to be indigenous (Cox and Possamai 2016: 6).

The indigenous population was deeply affected by European settlement which occurred rapidly after the British Explorer Captain James Cook claimed Australia for the British Crown in 1770. In his book, *A Place for Strangers*, Tony Swain (1993: 115) refers to the first 50 years after white settlement as "death as an ethnographic context." Swain reports that by 1850 in New South Wales and Victoria, just 4% (10,000 people) of the original population had survived. In addition, the Aboriginal population was displaced, removed from their homelands and forced to live in areas that completely changed their traditional links to the landscape. Swain (1993: 116) describes these conditions:

> Intensive pastoral activity spread, using primarily convict labour. Under these conditions, Aborigines were compelled to leave their homelands and form new social aggregates on the outskirts of towns, on the few benign stations, or on mission settlements. Those refusing to comply were visited by punitive expeditions . . . which at times [resulted in] murdering every Aborigine in a district.

What we know about indigenous religions in Australia prior to extensive contact with the white population is derived primarily from the research of 19th-century ethnographers and from the reports of missionaries. Much academic interest in the Aboriginal people was fuelled by evolutionary theories that described the indigenous groups of Australia as primitive, backward and at the lowest level of human development. Writing as late as 1938, the anthropologist A.P. Elkin asked: "Are the Aborigines the lowest race of mankind?" (1964 [1938]: 23). Elkin responded that this question is "not easily answered," but pointed to "some primitive features" as evidence that might support the contention, such as "the thickness of the skullcap, the shape of the face, the retreating nature of the forehead and the comparatively small brain" (*ibid.*).

The remote desert region of central Australia, which remained largely unaffected by colonial and Christian influences until relatively late in the 19th century, became the focus of intense scholarly interest among early researchers to the region.[1] The relative isolation of the populations convinced evolutionists that by studying the primitive "tribes" of the area as "living fossils," we could gain insight into the way humans from the Stone Age had lived and thought and on what basis they had formed rudimentary social structures (Charlesworth 2005: 6). In 1894, William Horn, a wealthy businessman, financed the first scientific expedition to the central desert regions of Australia. A number of eminent scientists, accompanied by prospectors, cameleers and Aboriginal guides, examined and collected specimens in a range of scientific fields, including the new academic subject called anthropology (Cox 2014a: 14). Baldwin Spencer from the University of Melbourne, who was the expedition's zoologist, edited the published reports of what came to be known as "The Horn Expedition" (Spencer 1896). While on the expedition, Spencer met F.J. Gillen, the Alice Springs Postmaster, who had good relations with a number of Aboriginal communities, although his knowledge of the local languages was limited. Following the Horn Expedition, Spencer returned to the central desert and collaborated with Gillen on a number of publications, the most important being *The Native Tribes of Central Australia* (1897). Some years after Gillen died in 1912, Spencer went back to the region to test some of his theories and published a revised version of the earlier book under the title *The Arunta: A Study of a Stone Age People* (1927), the opening pages of which display his evolutionary assumptions:

> Just as the platypus, laying its eggs and feebly suckling its young, reveals a mammal in the making, so does the Aboriginal show us, at least in broad outline, what early man must have been like before he learned to read and write, domesticate animals, cultivate crops and use a metal tool.
>
> *(Spencer and Gillen 1927: vii)*

Some years later, T.G.H. Strehlow, who had grown up among the Arrernte on the Hermannsburg Lutheran mission and knew the indigenous languages of the region almost from birth, evaluated the principal weakness of Spencer and Gillen as "having been forced by their ignorance of the aboriginal languages to omit texts of the songs sung" during the various festivals they observed and which they described in their books and articles. Strehlow concluded: "This was obviously quite unsatisfactory" (Strehlow 1971: xv–xvi).

From Spencer and Gillen in particular the idea was derived that Aboriginal people's religious perspectives were defined by what they called "The Dream Time," and that their society was organized around totemic symbols and thus could be summarized under the classification "totemism" (Moore 2016: 87–88; Charlesworth 2005: 7). These two concepts had a lasting effect on how Aboriginal peoples were described in later literature and provided the basis for highly influential theoretical models, such as Émile Durkheim's notion of totemism as the most elementary form of religious life (Durkheim 1915: 92–96; Charlesworth 2005: 7) and Sigmund Freud's conjectures about totemism being closely allied with the universal primitive taboo against incest, which in turn was integrally connected to his theory of the Oedipus complex (Freud 1938 [1919]: 15–36; Evans-Pritchard 1965: 42). The notion "Dream Time," later simply called "The Dreaming" by the noted anthropologist W.E.H. Stanner (2009 [1956]: 57–72), and its connected concept "totemism" thereby became widely promulgated as the defining characteristics of the indigenous religions of Australian peoples generally. Since "Dreaming" and "totemism" are still generally considered nearly universal concepts found among traditional Australian indigenous peoples, I will consider these two key terms in detail (Cowan 1992: 39–52; 82–95).

The Dreaming and totemism

W.E.H. Stanner, who began studying the Murinbata people in the far north of the Northern Territory in the 1930s, continued field studies there for the next 30 years. It is largely on the basis of this research that he constructed his theory of Aboriginal "Dreaming." In an essay first published in 1956, Stanner (2009 [1956]: 57) claimed that "a central meaning of The Dreaming is that of a sacred, heroic time long ago when man and nature came to be as they are." He differentiated the indigenous understanding of time as conveyed in the idea of Dreaming from Western concepts of time explaining that he had "never been able to discover any Aboriginal word for *time* as an abstract concept" (*ibid.*, emphasis in original). He added, "the sense of 'history' is totally alien here" (*ibid.*).

Stanner described The Dreaming as referring to "many things in one" (2009 [1956]: 58). He called it "a kind of narrative of things that once happened" and yet suggested that it is not restricted to the past because it is a "charter of things that still happen" (*ibid.*). In this sense, The Dreaming constitutes a "principle of order transcending everything significant for the Aboriginal man" (*ibid.*). Stanner identified three principal elements in stories or myths about '"The Dreaming." The first he called "great marvels" which describe

> how all fire and water in the world were stolen and recaptured; how men made a mistake over sorcery and now have to die from it; how the hills, rivers, and waterholes were made; how the sun, moon and stars were set upon their courses; and many other dramas of this kind.
>
> *(2009 [1956]: 61)*

The second element explains how social order was instituted for the first time:

> How animals and men diverged from a joint stock that was neither one nor the other; how the black nosed kangaroo got his black nose and the porcupine his quills; how such social divisions as tribes, clans, and language

> groups were set up; how spirit-children were first placed in the waterholes, the winds, and leaves of trees.
>
> (*ibid.*)

Stanner's third element demonstrates the close connection between The Dreaming and the present by connecting The Dreaming to the Aboriginal idea of law. He explains that "many of the main institutions of present-day life were *already ruling* in The Dreaming, e.g., marriage, exogamy, sister-exchange, and initiation, as well as many of the breaches of custom"(*ibid.*). He adds that law does not mean "legal" or "legalistic," but the ways of life that describe Aboriginal social structures, rituals, and relationships to spiritual or unseen forces. Then, in an often quoted phrase, Stanner described The Dreaming, not as everywhere (spatially) or at all times (temporally), but as both at the same time, the meaning of which he encapsulated in the term "everywhen" (2009 [1956]: 58).

The idea of The Dreaming or Dream Time, which has become commonly associated in popular circles with a pan-Aboriginal spiritual worldview, has been criticized by Tony Swain as too simplistic and thus prone to misunderstanding. In *A Place for Strangers* (1993: 16), Swain argues that The Dream Time, as first introduced by Spencer and Gillen, inappropriately imposed Western concepts of time, both linear and cyclical, on Aboriginal worldviews. In order to overcome the distortions created by Western ideas of time, Swain constructed two fundamental interpretative categories: rhythmed events and abiding events.

By rhythmed events, Swain (1993: 19) means that "in Aboriginal thought there is nothing beyond events themselves." Events simply occur. They are observed and are rhythmic and predictable. Examples provided by Swain include: the Milky Way stretched out across the centre of the sky, bandicoots backing into their burrows, light glimmering, the outline of trees and objects as clearly defined, shadows as variegated and the sun sinking (1993: 19). These concrete events occur regularly and demonstrate how time is thought of in pragmatic terms rather than as an abstract category.

The concept of "abiding events" points to the fact that the true meaning behind The Dreaming is "not temporal but spatial" (Swain 1993: 22). An abiding event is secure because it resides in a place. In this sense, abiding events and rhythmic events are coterminous, linked not through time but place. This explains the fundamental importance of the landscape on which sacred places are dotted and about which dreamtime stories are told. For Swain, The Dreaming is made manifest in rhythmic and abiding events, both of which are witnessed in the landscape which embodies the stories of the totemic ancestors crossing the land, making tracks and then by re-entering the land, making the particular places where they settled sacred (1993: 23).

This then brings us to the second key element in Aboriginal beliefs, that of the totem. A.P. Elkin described The Dreaming as being closely associated with a person's "cult totem." He explained: "To ascertain a man's cult totem . . . the best way is to ask him what is his 'dreaming'" (Elkin 1964 [1938]: 155). The answer provided may simply refer to an animal or plant (the totem), or the person may describe a myth "narrating the doings and travels of a great hero or heroes in the mythical age, 'the eternal dreamtime', a time which is past and yet, in a sense present" (*ibid.*). The symbol of The Dreaming, according to Elkin, was the churinga (*tjurunga*), a stick or object with secret symbols painted or carved on it: "When the newly initiated man is allowed for the first time to see and handle these objects and is rubbed with them, he is brought into conscious touch with this eternal dream-time" (Elkin 1964 [1938]: 187). The connection between the "eternal dream-time" and the totem is made evident by the central importance of the churinga as a symbol of a particular animal or plant. Elkin explains: "The life or spirit part of a totemic species is associated with a special churinga of wood or stone" (*ibid.*).

One of the most insightful and informed explanations of how the totemic ancestor is understood among the Arrernte (Aranda) of the Australian central desert region has been written by T.G.H. Strehlow in his seminal book *Aranda Traditions* (1947) and in his later booklet, "Central Australian Religion" (1978). Based on his extensive research among the Arrernte from the early 1930s until the 1960s, Strehlow compiled extensive accounts of Arrernte myths of creation. In most versions, Arrernte myths describe the Earth originally as "a featureless, desolate plain" (Strehlow 1978: 14). Below the surface of the Earth, life in the form of "thousands of uncreated supernatural beings" already existed in a state of slumber (Strehlow 1978: 15). Life as we know it began when the supernatural beings roused from their sleep and, according to Strehlow, "broke through to the surface of the earth; and their 'birthplaces' became the first sites on the earth to be impregnated with their life and power" (*ibid.*). These original and eternal beings are what are now called totemic ancestors.

Most totemic ancestors appeared in the form of animals, such as kangaroos, emus, bandicoots, honey ants and many other species. They wandered along the surface of the landscape, and as they travelled, they created the present geographical features visible in the landscape of the central desert regions, including the mountains, hills, plains, and water features such as swamps and springs. In the wide region inhabited by the Arrernte, Strehlow explains,

> there was not a single striking physical feature which was not associated with an episode in one of the many sacred myths, or with a verse in one of the many sacred songs, in which aboriginal religious beliefs found their expression.
>
> *(Strehlow 1978: 16)*

The songs, which formed part of the ceremonial life of the Arrernte at the time of Strehlow's research, were believed to have been composed by the totemic ancestors themselves.

In their wanderings across the surface of the Earth, the totemic ancestors left what Strehlow calls "a trail of 'life' behind them," which has a direct bearing on how the ancestors influence and relate to their descendants today (Strehlow 1978: 20). Strehlow explains: "According to reincarnation beliefs, some part of this [trail of life] could enter into the body of a human mother who crossed these trails, and could take on a fresh existence as a human infant" (*ibid.*). It is at the place where a woman first became aware that she was pregnant that the child she was carrying assumed the totem of the ancestor associated with that specific location on the landscape. This does not mean that Aboriginal people were ignorant of how conception occurred, but, according to Strehlow, in traditional belief, every person is comprised of two souls: one the result of normal intercourse between a man and a woman, and the other "the 'life' of one of the immortal supernatural ancestors which entered the body of an already pregnant woman at some definite point in the landscape" (Strehlow 1978: 21). When an Aboriginal person says, for example, "I am a kangaroo man," Strehlow suggests that what was meant was that "he had been reincarnated at a certain place from the trail of life left there by a supernatural being from whose body kangaroos had also originated at the beginning of time" (Strehlow 1978: 25).

It is important to note that if we follow Strehlow's interpretation, one's totemic ancestor is not determined by kinship or lineage, but by the "conception site," as noted previously, the place where the mother first became aware she was pregnant. This means that every individual possesses a personal totemic symbol, what Strehlow refers to as "personal monototemism." He explained that personal monototemism means that "every individual had only one personal totem; and it was this personal totem that mainly determined the nature of his rights, duties, and functions in the religious sphere" (Strehlow 1978: 26). The fact that individuals possessed their own personal totemic ancestors meant that members of the same kinship group normally had different totemic ancestors, and any particular sacred ceremonial site, called by Strehlow *pmara kutata*, linked together those that shared the same totemic ancestor.

This would suggest that personal monototemism overrides kinship identities, but this is not the case. Strehlow explains that personal monototemism occurs within "polytotemic communities" that are determined primarily, although not exclusively, along kinship lines. The key words associated with polytotemic communities are *pmara kutata*, the ceremonial site where the totemic ancestors were believed to have emerged at the beginning of time and where they returned; and the *njinaŋa* section, which, in Strehlow's words "includes all men, women, and children of a given totemic clan who stand to one another in the relation of fathers, sons, brothers, sisters, and daughters, relationship being determined both by actual kinship and by class ties" (Strehlow 1947: 139).

Strehlow provides the example of the honey ant totem. He notes that when a major ceremony was held at the pmara kutata of "the mythical birthplace of hundreds of ancestral honey-ants," all males of the njinaŋa section, which was organized along kinship lines, were required to be present and "were expected to contribute also at least one or two of their own totemic acts" (Strehlow 1978: 42). Members of the honey-ant totem who were not from the section were permitted to attend, but they ranked below the men of those who belonged to the ŋjinaŋa. In this way, personal monototemism was united into a polytotemic community, demonstrating the primary significance of kinship in determining societal organization.

We see from these brief accounts of two key concepts found within traditional Aboriginal religions, The Dreaming and totemism, that Aboriginal beliefs, practices, ceremonial life and the resulting social order were complex, intricate and involved sophisticated ways of thinking. The central place of the landscape confirms, as noted by Swain, that Dreaming needs to be thought of in spatial rather than temporal terms, or at the very least as a combination of the two as expressed in Stanner's notion "everywhen." Kinship relations also relate to spatial location, and, as Strehlow has shown among the Arrernte, group identity according to ancestral traditions prevails, even when individual associations with non-lineage related ancestors play an important function linking the landscape to personal identity.

The Māori of New Zealand

We turn now to consider the religion of the indigenous peoples of Aotearoa, which is the indigenous Māori name for New Zealand, and is usually translated as "long white cloud" or "land of long white cloud" (McLintock 1966: www.tearagovt.nz/en/1966/aotearoa, accessed 18 July 2016). The precise date of the first human settlement of New Zealand is a matter of debate, but current understanding is that the first arrivals came from East Polynesia as late as the 13th century CE, although other scholars place it earlier and some refer to two waves of immigration from Polynesian regions (Craig 2004: 176–177). According to the *Encyclopedia of New Zealand*, the Polynesian settlers did not identify themselves with a collective name until the arrival of Europeans after 1642, when they adopted the name Māori, meaning "ordinary," to distinguish themselves from the whites (*Te Ara. The Encyclopedia of New Zealand*, www.teara.govt.nz/en/history/page-1, accessed 18 July 2016).

In describing the Māori indigenous religion, it is important to note that in the 2001 census, 98% of Māori who stated a religious affiliation were Christian (www.stats.govt.nz/browse_for_stats/people_and_communities/maori/census-snapshot-maori.aspx, accessed 18 July 2016). Writing in 1980, New Zealand scholars of religion Brian Colless and Brian Donovon (1980: 10) called New Zealand "basically a Christian nation." Recent statistics demonstrate a changing picture. In the 2013 census, the number of Māori who claimed to be Christian had declined by 19.2%. Of the total population in New Zealand in the 2013 census, 41.9% stated "no religion" – an increase of 7.3% from the 2006 census (www.stats.govt.nz/Census/2013-census/profile-and-summary-reports/quickstats-culture-identity/religion.aspx, accessed 18 July 2016). Clearly, New Zealand is becoming more secular, and the Māori are no exception to the general trend. Much of what I describe about Māori religion and cosmology needs to be understood in the light of contemporary changes, including the fact that the ways in which the tradition is preserved may have more to do with cultural identity and Māori pride than with the practice of ancient religious traditions.

Early sources about Māori creation myths are recorded by scholars writing towards the end of the 19th century and the beginning of the 20th century. Two important figures are Elsdon Best, who has been described "as New Zealand's foremost ethnographer" (Bennington 2005; Cox 2014a: 36); and S. Percy Smith, one of the founders of the Polynesian Society and editor of the *Journal of the Polynesian Society* from 1901 until his death in 1922. Both Best and Smith relied for their ideas about Māori cosmology on information supplied by Te Whatahoro, otherwise known as John Alfred Jury, who was born in 1841 of an English father and a Māori mother. Te Whatahoro worked as an interpreter in the Native Land Court, and for over 40 years, produced voluminous notebooks purporting to preserve indigenous traditions (Holman 2010: 224). During the 1960s, he attended sessions in a *whare-wānanga*, which were secret, upper-class Māori religious societies, where he claimed to have learned ancient traditions directly from two *tohunga* (indigenous religious practitioners). In 1900, Te Whatahoro was baptized into the Church of Jesus Christ of Latter Day Saints (Mormons). Thirteen years later, S. Percy Smith published a translation in English of a two-volume work written by Te Whatahoro under the title *The Lore of the Whare-Wānanga*, which outlines in detail the alleged ancient Māori belief in a supreme being called Io (Te Whatahoro 1913, 1915). More recently, a summary of Māori religious beliefs was written by the New Zealand scholar James Irwin (1984). Other recent important contributions offering details of pre-Christian Māori beliefs and practices have been written by Michael Shirres (1997), a Roman Catholic theologian working in New Zealand and Māori Marsden (2003), an indigenous Anglican priest.

Three versions of the creation myths can be found in the literature: one which emphasizes creation by Io, the postulated Māori supreme being; a second, a widely known and highly popular version that attributes creation to a result of separating the primeval parents Rangi (the sky father) and Papa-tua-nuku (the earth mother); and a third that focuses on creation by the *atua*, sometimes called gods, departmental gods or ancestors with ongoing influence over particular domains of life (Irwin 1984: 81; Knappert 1995: 22–23). These versions are all inter-related and differ mainly according to the emphases they place on the supernatural beings described.

The creation by Io is controversial, since a debate centres on whether or not Io was introduced into Māori cosmology after contact with Christianity. Best and Smith were the principal promoters of the primordial belief in Io as a Creator God, but their source, as we noted, was Te Whatahoro, who between 1886 and 1888, assisted Mormon elders in translating the Book of Mormon into the Māori language (Parsons 2010). Many of the ideas contained in *The Lore of the Whare-Wānanga* bear uncanny resemblances to Mormon stories about the discovery of the tablets that contained the Book of Mormon in the hills of New

York by Joseph Smith (Bowman 2012: 14). *The Lore of the Whare-Wānanga* also contains accounts of creation stories that resemble Mormon doctrines of creation as emanations of an eternal, material God (O'Dea 1957: 120). The Christian theologians Michael Shirres and Māori Marsden claim that belief in Io demonstrates the high intellectual ability and theological awareness of the Māori prior to Christian influences. The way they describe creation, however, sounds in many ways like a recounting of Christian theological themes. Shirres (1997: 114) argues that Io existed at the beginning "as the life force which is the source of our life as conscious, knowing and willing beings." Io existed in the "realm of potential being" and encompassed both "active" and "passive" elements in one being (Shirres 1997: 33). Out of a conjunction of the active and passive elements, Io initiated creation through a process of "genealogical recitation" or "naming." During a period of 27 nights, Io first created the void, the abyss and the night. Then, out of these, Io caused existence, first as a seed, then an ever-growing plant, eventually developing into consciousness, life, time and space, sense perception and matter emerging in Rangi (the sky) and Papa-tua-nuku (the earth). Io delegated the further process of creation to Rangi and Papa (Shirres 1997: 25–29).

A controversy about the pre-Christian belief in Io has continued to the present day with scholars debating both sides of the case. Both Jonathan Z. Smith, the eminent historian of religions at the University of Chicago, and Peter Buck (Te Rangi Hīroa), who was a highly respected Māori and Pacific scholar, argue that Io as a name for a Māori supreme being with attributes similar to the Christian God is a recent development (Smith 1982: 68–80; Te Rangi Hīroa 1950; Cox 2014a: 53–58). Shirres and Marsden, on the other hand, contend that Io represents an ancient Māori belief in a high god (Cox 2014a: 44–51). Several convincing reasons exist for adopting a sceptical attitude towards the sources on which scholars rely in support of the antiquity of Io, including the very close relationship between the way Io is depicted and Christian theological concepts, and the fact that the primary source for the Io belief was based on the evidence provided by the Mormon convert Te Whatahoro.

The story of creation that appears closer to Māori traditional cultural roots and also shares ideas with other Polynesian creation stories is the second version, which describes creation occurring by the separation of Rangi and Papa. According to this account, in the beginning, Rangi (sky) and Papa (earth) clung so tightly to each other that they blocked out all light, causing their children to live in perpetual darkness. The children were dissatisfied with this, so they decided to separate their parents. One child (Tū-mata-uenga) wanted to kill them, but his brother (Tāne-mahuta) sought only to disengage them. All Tāne's brothers sought to disentangle their parents, but failed. Finally, Tāne succeeded in dislodging them forcibly. Following their separation, Rangi and Papa grieved for each other. Rangi's sorrow was expressed in rain (tears) and Papa's sadness was manifested as tears in the form of the rising mist. One of the brothers (Tāwhiri-Mātea) disagreed with the separation of Rangi and Papa, and in association with his father Rangi, made war against the brothers. Tāwhiri-Matea became the god of the winds and hurricanes, by which he periodically destroyed the forests. Tāne, who had separated the primeval parents, became the god of the forests and the birds. Two of the brothers, the "father of fish" and the "father of reptiles" separated from one another – one going to the sea and the other staying on land. Other brothers became gods of other parts of the world and life: war, weather, agriculture, peace, the god of dominion over human beings, the god of evil and death, and the god of light and life (Reilly 2004: 3–5).

One interpretation of the mythological stories of Rangi and Papa, their offspring and their warfare is provided by James Irwin in his *Introduction to Maori Religion*. Irwin (1984: 14) suggests that this version of creation provides "explanations of the opposition between sky and earth, the struggle for existence in a hostile environment, and a rationale for the continuing rivalry among human beings." Another scholar of Pacific religions, Michael P.J. Reilly (2004: 5), argued that the separation of Rangi from Papa "initiates a process of differentiation whereby the parents, their various sons, and their descendants, become associated with aspects of the natural world of the Māori."

The creation of the natural world by the descendants of Rangi and Papa leads to the third version of creation, that by atua, who, as the children of Rangi and Papa, assumed responsibility for various aspects of the natural world and human social ordering. In this sense, they operate as autochthonous ancestors, establishing the world as it is now experienced and founding human relations. Reilly (2004: 5) explains: "With the parental separation successfully completed, the world of the *atua* unfolds and expands, and takes on its present form, with subsequent generations continuing a process of differentiation as they carve out their separate lives within it." It is important to note that in most Christian churches throughout New Zealand, the indigenous word for God is Atua, pointing towards the central role of lineage in Māori traditional society. This observation is underscored by Reilly (2004: 12):

> By narrating, reciting or singing these creation accounts, Māori *tohunga* (priests) bound together the immanent possibilities found within them, just as a tribal genealogy has the flexibility to allow first one, then another descent line to be emphasized, while still holding the larger web of connections in place as part of the total kinship system.

Māori understanding of humanity and society

Māori Marsden and James Irwin have written substantially on the practical ideas found in Māori religious understandings, which continue to influence, even in Christian guises, how the Māori relate to the other humans and to society. According to Irwin (1984: 21) and Marsden (2003: 6), in Māori anthropology, each person is born with a life principle, called *mauri*. This is imparted at birth and refers to the "personhood" or "soul" of the individual (Irwin 1984: 21). Closely related to the concept of mauri is the central idea of *mana*, which is obtained in birth rituals recited by the infant's father or by the tohunga (*ibid*.). Mana was discussed widely in early anthropological literature on Pacific island peoples, initially by the missionary anthropologist R.H. Codrington who worked in Melanesia. In his book *The Melanesians*, Codrington (1891: 118) described mana as "a power or influence, not physical, and in a way supernatural." The force Codrington identified as mana, is believed among the Māori to reside not only in humans, but also can be found in impersonal objects or even understood as prestige, status or authority that, over time, can be diminished or lost altogether (Irwin 1984: 21). A person's mana determines his or her success in life endeavours and, as such, requires protective devices to insure that it is maintained.

Another important Māori concept that has been studied widely is *tapu*, or taboo. It refers to a state of being set apart (Knappert 1995: 286). It is commonly thought of as a prohibition against some behaviour, but its primary function is to provide protection against the diminution of mana. If a tapu is broken, it is not the tapu or prohibition that wreaks harm, but the mana that has become uncontrolled. In this sense, according to Māori Marsden (2003: 5), tapu refers to a person or object that is "removed from the sphere of the profane and put into the sphere of the sacred," thereby making it "untouchable." For example, traditionally, no commoner would ever dare drink from the cup of a chief (it is tapu); it was tapu because the cup was imbued with the mana of the chief (Irwin 1984: 24).

A key idea in Māori understanding of social relations is *noa*, a term closely related to the rituals that provide purification from the results of a tapu that has been breached. In fact, the opposite to tapu is noa, since noa is translated as "common" in contrast to "sacred" (a thing set apart) (Irwin 1984: 28). Rituals associated with making a person or a thing noa are termed *whakanoa* (to cause to be ordinary) (Irwin 1984: 29). Such rituals aim at redressing the consequences of a breach of tapu so that mana is not diminished or lost, nor a person's mauri destroyed by the invasion of malignant spirits that are released when a tapu has been violated.

Different whakanoa rituals were performed as proportionate responses to the gravity of the offence committed. For a minor violation of tapu, it was often sufficient for the offender to touch a piece of cooked meat, since cooked meat – as noa – was believed to neutralize the effect of breaching the tapu. Another remedy was to wash one's hands in water, since according to tradition, running water was always regarded as noa (Irwin 1984: 29). More serious infractions of tapu required the person to visit a sacred place where a tohunga would perform necessary rituals, often at the site of a running stream, or on other occasions beside a fire. James Irwin (1984: 30) explains: "In all cases the ritual required *whakihara* (confession), recital of the ritual *karakia* [sacred verses] with some appropriate action, and finally a declaration by the *tohunga* of the effectiveness of the purification" (See also, Marsden 2003: 7–8).

Cases involving the deliberate breaking of a tapu were considered to be a challenge to the mana of the atua and thus required the intervention of a tohunga with great mana. Māori Marsden (2003: 7–8) provides examples of whakanoa rituals. Marsden explains that different types of food, ferns or other herbs were cooked and, after they became cold, were placed on the person's head, which was considered "the most sacred part of the body" (Marsden 2003: 7). It was believed that by cooking the food or herbs, the mauri (life principle or soul) of the plant was "released," and then, by placing the cooked food or herbs on the head of the offender, the individual was made noa (common), thereby neutralizing the mana released by the deliberate violation of the tapu (*ibid.*).

A word about the tohunga or traditional religious practitioner will be helpful at this point. Maori Marsden (2003: 14) argues that tohunga often is translated as "expert," but he claims that this is an incorrect application. In actuality, he explains, the word is derived from the stem "tohu," which as a verb means a sign or a manifestation. Tohunga thus means "a chosen one" or an "appointed one." Marsden defines a tohunga as "a person chosen or appointed by the gods to be their representative and the agent by which they manifested their operations in the natural world by signs of power (*tohu mana*)" (*ibid.*). Jan Knappert (1995: 301) likens the tohunga to a traditional shaman, in that the tohunga understands the "language" of the spirits, communicates with them and through this obtains knowledge of "the past, present and future."

As a shamanistic type of religious practitioner, the primary function of the tohunga was to mediate between the atua and the people in order to ensure the welfare of the community. The tohunga did this by giving advice on the best time to engage in certain activities, such as where the most productive fishing grounds were at certain seasons, or what the most effective methods were for ensuring success in economic activities. They were also experts in propitiating the atua by performing various ceremonies and were also versed in the sacred stories, traditions and genealogies of the people to whom they belonged (Marsden 2003: 14–15).

The continuing influence of kinship and location

The continuing influence of traditional Māori religion is seen most clearly in the rituals of death. According to James Irwin (1984: 49), the funeral ritual (*Tangihanga*) takes its name from the verb *tangi*, to weep, and, he claims, in modern times it "is the most familiar of all the ancient rituals." Irwin (1984: 49) notes that although today, most of those participating would describe themselves as Christians, the essential elements observed are those obtained from the indigenous religion. In the traditional view of death, there was always a spiritual cause for death in addition to natural causes. As a result, traditionally, every death had to be requited, or paid for (*utu*), "to redress a wrong either in blood or plunder" (Irwin 1984: 50). Requital for death was thus a central concept in indigenous Māori religion. Nowadays, Irwin explains, utu is paid in three main ways: a) by weeping and wailing; b) by oratory, sometimes taunting another person who must jeer back to the original speaker; and c) by exudings from the nose and mouth (*ibid.*).

Evidence of the persistent influence of Māori indigenous religions in contemporary times is found in the widespread importance of the *marae*. The current website of "100% Pure New Zealand"[2] defines the marae as "a fenced-in complex of carved buildings and grounds that belongs to a particular *iwi* (tribe), *hapū* (sub tribe) or whānau (family)" (www.newzealand.com/au/feature/marae-maori-meeting-grounds/, accessed 18 July 2016). The website adds: "Māori people see their marae as *tūrangawaewae* – their place to stand and belong" (*ibid.*). Jan Knappert (1995: 181) explains that in traditional Polynesian societies, the marae was "a public square in front of the temple or chief's mansion" and normally "was enclosed by stone or coral rock walls." Nowadays, the marae is used for meetings, celebrations, funerals, educational workshops and other important community events. The most important of the buildings within the marae is the *wharenui*, or carved meeting house. A wharenui resembles the human body in structure, and usually represents a particular ancestor of the iwi. The *tekoteko* (carved figure) on the roof top in front of the house represents the head, and the *maihi* (front barge boards) are the arms held out in welcome to visitors. The *amo* are short boards at the front of the *wharenui* representing legs, while the *tahuhu* (ridge pole), a large beam running down the length of the roof, represents the spine. The *heke* (rafters), reaching from the *tahuhu* to the *poupou* (carved figures) around the walls, represent the ribs. Many wharenui contain intricate carvings and panels that refer to the *whakapapa* (genealogy) of the tribe, and to Māori stories and legends (www.newzealand.com/au/feature/marae-maori-meeting-grounds/, accessed 18 July 2016). Leaders of each marae were traditionally determined according to genealogy, but more recently, these can also be assigned according to what is called "situational identity" based on communally accepted association rather than direct lineage (www.awataha.co.nz/About+Us/What+is+a+Marae.html, accessed 18 July 2016).

Frequently, death rituals are held at the marae in the wharenui. As death becomes imminent, or immediately after death, a karakia is recited to release the person's spirit from the body and to prepare for the journey on the road to the spirit world. According to tradition, the spirit of the person journeys to a rocky outcropping on the northern tip of the North Island, Cape Reinga. From there, the spirit descends to the floor of the ocean to begin the pilgrimage that will take it to the abode of Hinenuiotepo, the guardian of the entrance to the land of the dead. She acts as a guardian to guide the spirits of the dead past the dreaded Whiro – the god of evil and destruction. The spirit's journey does not begin until the funerary rituals are completed, otherwise it will become a wandering spirit that is troublesome to the family. The funeral rituals, as described here, are punctuated with much weeping and mourning (Irwin 1984: 51–54).

The myths of creation featuring Io, Rangi and Papa and the atua; the key concepts including the life principle mauri, mana, tapu and noa; and the views of death and the symbols of the tradition at the marae all form what James Irwin (1984: 5–20) calls the "Māori worldview." This, he contends, still informs the way the Māori think about reality. I would suggest that Io conceived as the Māori version of a creator god is a post-Christian invention that has been inserted into Māori cosmology by Christian missionaries and by sympathetic ethnographers, such as Elsdon Best and S. Percy Smith. It is much more likely that in their ancient form, Māori stories of creation emphasized the role of the atua, who are both the creators of the natural order and the autochthonous ancestors of all Māori groups. The importance of location is evident even today as represented by the pivotal importance of the marae as a meeting ground for the iwi (kinship-orientated local groups). That the wharenui is built in the shape of a human body with an ancestor at the pinnacle and is adorned with drawings depicting genealogical stories confirms the persistent importance of kinship within a particular locality despite the overwhelming influence of Christianity within contemporary Māori society.

South Pacific Island religions

The final section of this chapter explores the indigenous religions of the South Pacific Islands, from Melanesia in the west, to Samoa, Tonga, the Cook Islands, the Society Islands and the Marquesas in the east. Robert D. Craig (2004: 2–6) argues that the settlement of central and eastern Polynesia began around 2,000 BCE, with migrations eastward from Melanesia to Western Polynesia arriving in Tonga by 1,300 BCE and Samoa by 1,000 BCE. Over a period of time, the languages and cultures of central Polynesia differentiated themselves from that of the western islands. Eventually, the more eastern and southern islands were settled with the Marquesas being reached around 400 CE, the Society Islands approximately 800 CE and New Zealand around 1,100 CE. He concludes that by 1,100 CE "essentially the last major regions of Polynesia had been settled" (Craig 2004: 5).

Melanesia, a term meaning "black islands" according to Swain and Trompf (1995: 4), was used first in the early 19th century by the French explorer Jules Dumont d'Urville, who applied it to the "crinkly haired black people" who populated New Guinea, the Solomon Islands, the New Hebrides and Fiji. In his important book, *Melanesian Religion*, Trompf (1991: 10) admits that to make generalizations about Melanesian religions is very difficult, but it is even more problematic to expand beyond Melanesia to include the islands of western, central and eastern Polynesia, and then to extend this further to the so-called Polynesian Triangle, which stretches north to Hawaii and south to New Zealand. Nonetheless, because the movement of peoples was most likely from west to east in this region, certain similarities occur that can lead to some broad generalizations about the cosmos, mythology and, in particular, the place of ancestors. Trompf (1991: 123) confirms this when he notes: "As the general cultural boundaries of the regions are relatively fluid . . . one should expect some overlapping of religious configurations."

The importance of the sky as a source of religious symbolism can generally be found across the regions. For example, Trompf (1991: 13) refers to the importance among Melanesian peoples of "moving objects in the sky – the sun and moon . . . and the Morning Star – all of whom often make their appearance in mythologies in human guise." Just as we saw with the Māori, accounts of early missionaries and ethnographers of the religions of central and eastern Polynesia refer to Io, a name Jan Knappert (1995: 129) calls "one of the Polynesian words for *God*, the Supreme Being" (emphasis in original). Knappert reports that as far north as Hawaii, Io meant hawk or falcon, suggesting that Io "was once a sun-god, since the falcon was a universal symbol – or an identification – for the Sun God" (*ibid.*). The association of Io as a sky or sun deity with the Christian idea of God appears to have occurred, just as it did in New Zealand, following contact with Christian missionaries. Craig (2004: 40) asserts that the "less known" story of creation among Polynesian peoples "is the creation by a single, supreme being who creates the world out of himself" (40). Craig expresses the general scepticism of the idea of Io as equivalent to the Christian God – which I mentioned with respect to the alleged ancient Māori belief in Io – when he observes: "Some scholars believe it came into being only after the introduction of Judeo-Christian beliefs among the Polynesians in the late eighteenth and early nineteenth centuries" (*ibid.*).

Within the Pacific Island regions, atua, meaning "the old one," refers, just as we saw with the Māori, to beliefs about the deities who are understood as ancestors of the people. Knappert (1995: 22) supports this interpretation when he defines an atua as "the spirit of an ancestor in Polynesia, who is revered like a god." According to Robert W. Williamson (1937: 9), in Samoa, atua were "the original gods" who dwelt in the heavens and "were believed to have been the progenitors of the other deities, and to have formed the earth and its inhabitants." The *Maori Dictionary of New Zealand* challenges the Christian underpinnings to this definition, at least when applied to the Māori, claiming that "this is a misconception of the real meaning," and explaining that "many Māori trace their ancestry from *atua* in their *whakapapa* and they are regarded as ancestors with influence over particular domains" (http://maoridictionary.co.nz/word/494, accessed 16 July 2016). Among the Tuamoto Islands of French Polynesia, also referred to as the Paumoto Islands, Williamson (1937: 26), citing the research of the French historian Eugène Caillot, describes atua as ancient chiefs who had been deified. This interpretation would confirm the connection between atua and ancestors. Trompf (1991: 129) reports that in Melanesia, mythic stories abound telling about "culture heroes" who "are frequently groups of Beings who pass through the land in mythic times, revealing to each tribes' Ancestors the skills of warfare and food production and the technologies for building, weaving etc."

Various deities are also reported by Williamson (1937: 16–23) among the Society Islands, in which the gods were ranked according to their responsibilities within corresponding areas among human communities. Those of "the highest order" included Tane, who, as we have seen, was regarded among the Māori as one of the children of the primeval parents Rangi and Papa, and who succeeded in separating the sky from the earth, thereby creating light. Williamson includes Tane alongside the highest ranking deities Tangaroa and Oro. He adds that research conducted by the missionary William Ellis between 1816 and 1822 suggests that Tangaroa was "generally spoken of by the Tahitians as the first and principal god, uncreated and existing from the beginning" in a way that sounds quite similar to how the Christian theologians Shirres and Marsden in New Zealand described Io, and again reflects a missionary interpretation that described indigenous societies as already having been prepared by God for the Christian message when it was delivered to them (Williamson 1937: 17).

What does seem to be reported consistently among the inhabitants of the South Pacific Islands are beliefs in what has been called "departmental gods" with various functions assigned to them (Williamson 1937: 8). In Melanesian contexts, Trompf (1991: 14) refers to "differing classes and hierarchies of spirit-powers." Craig (2004: 41) clarifies how the spiritual order was established. He asserts that Pacific Island myths often begin with night or darkness ("*pō*"), which begets light, after which the Earth as we know it begins to appear. The relation between darkness and light is frequently depicted as a sexual union, such as we see in stories of the Sky Father and Earth Mother. Craig concludes: "Creation by sexual union can be found in all of the island groups" (*ibid.*).

Some examples of creation stories provided by Craig will illustrate both the variations and similarities of cosmogonic myths and how these influenced the beliefs about the hierarchical ordering of spiritual beings among the South Pacific Island peoples. The first comes from Samoa. According to Craig's summary (2004: 45–46), the Samoan creation story features High Rocks (Papatu) which united with Earth Rocks (Papa'ele), producing the "original gods" – first the gods of rocks and earth, and then the god of the underworld. After seven generations, the god Tangaloa (also Tangaroa in other Polynesian contexts) was born, who became "the principal god, the progenitor of all other gods and humans on earth" (2004: 45). Craig relates that in one version of the myth, Tangaloa had a bird-son called Tuli, who flew down to the Earth but found nowhere on which to land. Tuli returned to Tangaloa and reported that the Earth had no solid land. Tangaloa then threw a stone down from the sky, and when it landed, it became firm earth. Other versions of this story are related in the literature, but all conclude with the creation of man and woman who become the progenitors of the human race.

Craig underscores the importance of a creation myth derived from the Cook Islands, principally the island of Mangaia, in which the origin of life is attributed to the female goddess Vari-ma-te-takere. According to Craig, the Mangaian story describes Vari-ma-te-takere in the beginning as living alone in the underworld. Out of her own body, she extracted six children who became in turn "the father of gods and men": the lord of the seas, the god of the birds, the female rock, the wind god and, finally, "a beloved daughter whom Vari held close to her" in the underworld. Vari's first-born son, called Vatea, "the great god of earthly creation," had visions of a beautiful woman, but failed to find her anywhere. He set a trap by putting a freshly ground coconut on the ground near a cave. When what Craig calls a "dainty arm" reached for the coconut, Vatea grabbed it and discovered that it belonged to the "beautiful goddess Papa," translated by Craig as "Foundation." Papa became the wife of Vatea. Their first-born sons were the twins Tangaroa and Rongo. Eventually, the twins argued and decided to divide creation between them. Tangaroa received everything coloured red while Rongo received the majority. As a result, Rango became the paramount god of Mangaia (Craig 2004: 47–48).

Craig concludes by suggesting that all Polynesian peoples construct a view of the world in which there exists a three-layered ordering: the upper level of the heavens, the dwelling of humans and an underworld. This conforms to the shape of a coconut. When a coconut is broken, Craig explains, it generally splits evenly, "leaving two level halves" (Craig 2004: 51). He likens this to the Polynesian view of the universe, with the top half of the coconut symbolizing the "dome of heaven with its stars, moons, sun, and clouds, and, of course the residing places of the deities" (*ibid.*). The lower half represents the underworld, which is also populated by various spirits. A plane or dividing line is imagined between the upper and lower sections, representing the location of human life, or the Earth's surface, where human communities are formed. Both the upper and lower worlds are divided into levels where the beings inhabiting these regions exist and are ranked according to their place in the creation stories and their continued influence within human societies.

According to Trompf (1991: 122), in Melanesia, the universe is not conceived in the tri-partite division Craig described with respect to the Polynesian Islands. This can be explained by the fact that "the majority of Melanesian cultures were land-locked, especially on the great island of New Guinea." This meant that most Melanesian peoples, because they were "spatially confined," conceived the cosmos "horizontally," resulting in the "consequent tendency for the habitations of spirit agencies, along with human settlements, to be located on (and be accessible from) 'ground level'" (*ibid.*). This does not mean that Melanesian societies did not possess sky myths, but they lived within circumscribed locations and regarded peoples who came from beyond the horizon as occupying "uncultivated or uncontrollable places" (Trompf 1991: 122). As such, local deities were conceived in terms of ancestors, whose names were known and who "minimized" unexpected or dangerous occurrences (*ibid.*).

In the cases of the South Pacific Island groups discussed, some conclusions can be drawn, although, as we noted both in Australia and New Zealand, local variations are highly significant. The principal role of the atua as departmental gods, autochthonous ancestors and as progenitors of the various groups among the Polynesian peoples appears a consistent theme. Doubt over the role of Io or the supreme being conceived in Christian terms is found among the accounts throughout Polynesia just as we found among the Māori. In Melanesia, Io does not appear, and clearly ancestors provide a central focus within a "horizontal" cosmology that differentiates one local group from another. The idea among all South Pacific Island peoples that the deities, as ancestors, relate to specific and identifiable social groupings, which are restricted to particular geographical locations, appears again and again.

Kinship and locality in an age of globalization: a conclusion

We have seen throughout this chapter that across the vast Oceanic regions from Australia through New Zealand to the South Pacific Islands, ancestors traditionally played a central role among each indigenous society considered. This appears to confirm the contention with which I began this chapter that indigenous religions are defined principally by two central characteristics: kinship and location. This conclusion has been disputed by some scholars, most recently by Bjørn Ola Tafjord of the University of Tromsø in Norway (2013: 221–243), who contends that this limited definition ignores or at least minimizes the dynamic character of indigenous religions by relying on descriptions that no longer apply in contemporary situations. Tafjord argues that to define indigenous religions as kinship-bound and limited to specific geographical regions perpetuates 19th-century assumptions that portray indigenous peoples as backward, primitive, limited and stuck in the past. Of course, we know now that such a stereotypical characterization of indigenous societies is grossly inaccurate. Nowadays, people living in even quite remote regions have mobile phones, computers with access to the internet and televisions with satellite connections – and they often travel to various parts of the world rapidly. Many indigenous people reside in countries far removed from their homelands, but they still consider themselves indigenous.

My response to Tafjord's objection is straightforward. The fact that we live in an age of globalization does not undermine the basic fact that indigenous societies are fundamentally constructed around kinship relations, nor does it contradict the claim that the religions of such societies are restricted cosmologically to myths and rituals relating to specific geographical locations. My argument does not deny the obvious observation that today, individuals who identify as indigenous actively respond to influences emanating from the outside world. Just as many indigenous religious adherents are both indigenous and Christian, rather than being either one or the other, so too do indigenous peoples live both according to their ancient traditions and engage with the forces of modernity. The fundamental issue concerning the central importance of kinship and location relates not so much to the ways that indigenous peoples respond to global stimuli, but to how traditional social authority is transmitted, exercised and legitimized. Competing authorities will be present in contemporary circumstances and varying authoritative structures will hold sway in different situations. Nonetheless, wherever the authoritative tradition based on appeals to a line of ancestors is maintained and continues to operate with community consent, indigenous religion is present, even if members of the community also acknowledge other authorities, such as the church or the national government. It is not even necessary that members of the community physically reside in the location to which they owe allegiance. They can still be bound by the authority of their ancestors while living in cities or in far-off countries, and continue to subordinate themselves to the traditions still operating in the place from which they originate. If, by contrast, the authoritative religious tradition is transformed into a religion that seeks converts, acts with missionary intent and minimizes the role of ancestral traditions, this no longer can be called an indigenous religion.

It follows from this that I am not suggesting that the indigenous religions of the vast Oceanic regions will eventually die away as they come increasingly into contact with global economic, political, scientific, educational and religious structures. I am convinced that the indigenous religions of these regions are highly adaptive and, although they often adopt practices from outside cultures, they can still maintain their allegiance to their ancestral traditions. Such allegiance is dynamic, rather than static, as has been documented in numerous studies of various movements in such societies that have taken aspects of other religions and made them their own. I have provided in another publication the example of the Rainbow Spirit Theology in Northern Queensland, Australia, which has adapted the ubiquitous indigenous symbol of the rainbow-serpent by making this equivalent to God the Father, and by promoting the radical idea that Jesus Christ is the incarnation of the rainbow-serpent (Cox 2014a: 89–111). We might also cite the case of Io, which some advocates of Māori culture, such as Māori Marsden, have reinterpreted as an expression of the Christian God, while at the same time advocating for the persistent value of a Māori view of the world (Marsden 2003: 2–23). Both the Rainbow Spirit Theology and a Māori-Christian version of Io fit into a category, following the influence of the Russian literary critic Mikhail Bakhtin (1981: 358–362), I have called "intentional hybridity" (2014a: 145–147). According to the theory of intentional hybridity, indigenous societies exercise power by constructing their own responses to globalizing forces, rather than, as they have often been depicted, passively succumbing to the hegemonic power of universal, proselytising religions.

A final point should be made in conclusion. Indigenous religions are important to study not primarily because they can be compared with religions that are based on universal cosmologies. Rather, indigenous religions should be respected as traditions in their own right and not, as they were in previous times, valued solely because they served as a preparation for one of the "world" religions (usually Christianity) (Cox 2014b: 196–198) or because they acted as some basic, primal foundation on which all religions are constructed (as proposed by framers of the "primal religions" theory) (Cox 2007: 9–31). Each indigenous religion should be studied in its own context, historically, socially and culturally, before tentative generalizations can be made about regional similarities, as I have done in this chapter in the case of the religions of Oceania, and before scholars construct comparative typologies that, subject to intense empirical scrutiny, are applied across geographical and cultural boundaries.

Glossary

Amo: among the Māori, a short board in front of the wharenui representing human legs.

Aotearoa: Māori name for New Zealand, generally translated as "land of long white cloud."

Atua: Māori traditional gods, departmental gods or ancestors. In wider Polynesian contexts, often translates as "old one."

Hapū: Māori word for sub-tribe.

Heke: among the Māori, wooden rafters in the wharenui, representing human ribs.

Iwi: Māori word for tribe.

Karakia: among the Māori, sacred verses or ritual incantations.

Maihi: among the Māori, front barge boards on a wharenui.

Mana: power or force that resides in humans and in impersonal objects, closely associated with prestige, status, success or authority.

Marae: fenced-in complex of carved buildings and grounds belonging to a particular tribe, sub-tribe or family, still common among the Māori.

Mauri: among the Māori, a word designating life principle, sometimes translated as "soul," that inhabits humans and natural objects.

Njinaŋa section: among the Arrernte, all members of a totemic clan determined by kinship relations.

Noa: Māori word translated as "common" or "ordinary"; the opposite of tapu.

Pmara kutata: among the Arrernte, a sacred ceremonial site where the totemic ancestors emerged at the beginning of time.

Pō: Polynesian word for night or darkness; associated with creation stories.

Poupou: among the Māori, carved figures placed around the walls on the inside of a wharenui.

Tahuhu: among the Māori, a ridge pole on a wharenui, representing the human spine.

Tangihanga: Māori death rituals; taken from the Māori verb meaning "to weep."

Tapu: among Polynesian societies, a state of being set apart; often thought of as a prohibition.

Tekoteko: among the Māori, a carved figure commonly placed on the rooftop of a wharenui.

Tjurunga,(also spelled churinga): among the Arrernte, a stone or wooden object with secret symbols painted on it giving it sacred significance.

Tohu mana: signs of power exercised by the Māori tohunga.

Tohunga: Māori indigenous religious practitioner; literally a "chosen one" or an "appointed one."

Tūrangawaewae: Māori phrase referring to a marae as a place to stand and belong.

Utu: a payment or a requital for the occurrence of a death in traditional Māori death rituals.

Whakanoa: among the Māori, rituals that make a person or thing noa; purification ritual.

Whakapapa: Māori word for genealogy.

Whakihara: among the Māori, a confession of a breach of tapu.

Whānau: Māori word for family.

Wharenui: Māori meeting house set within the grounds of the marae.

Whare-wānanga: secret, upper-class religious societies in Māori tradition; called "schools" or "houses" of learning.

Notes

1 The first mission station in the central desert was established by German Lutherans in 1877 among the Arrernte peoples at Hermannsburg around 125 kilometres southwest of the Alice Springs post and telegraph station.

2 The webpage for "100% Pure New Zealand" describes itself as "the official travel website for New Zealand."

Bibliography

100% New Zealand. 'Marae – Māori Meeting Grounds', www.newzealand.com/au/feature/marae-maori-meeting-grounds (accessed on 18 July 2016).

Awataha. 'What Is a Marae?', www.awataha.co.nz/About+Us/What+is+a+Marae.html (accessed on 18 July 2016).

Bakhtin, M.M. 1981. *The Dialogic Imagination: Four Essays*. Austin: University of Texas Press.

Bennington, S. 2005. 'Foreword to Reprint', in Elsdon Best (ed.), *Maori Religion and Mythology: Being an Account of the Cosmogony, Anthropogeny, Religious Beliefs and Rites, Magic and Folk Lore of the Maori Folk of New Zealand, Part I*, no page numbers. Wellington: Te Papa Press.

Bowman, M. 2012. *The Mormon People: The Making of an American Faith*. New York: Random House.

Charlesworth, M. 2005. 'Introduction', in Max Charlesworth, François Dussart and Howard Morphy (eds.), *Aboriginal Religions in Australia: An Anthology of Recent Writings*, pp. 1–27. Aldershot: Ashgate.

Codrington, R.H. 1891. *The Melanesians: Studies in Their Anthropological Folklore*. Oxford: Clarendon Press.

Colless, B. and P. Donovan. 1980. 'The Editors' Introduction: The Religion of the New Zealanders', in Brian Colless and Peter Donovan (eds.), *Religion in New Zealand Society*, pp. 9–14. Edinburgh: T. and T. Clark.

Cowan, J.G. 1992. *The Elements of the Aborigine Tradition*. Shaftsbury, Dorset and Rockport, MA: Element Books Ltd.

Cox, J.L. 2007. *From Primitive to Indigenous: The Academic Study of Indigenous Religions*. Aldershot: Ashgate.

Cox, James L. 2014a. *The Invention of God in Indigenous Societies*. London and New York: Routledge.

Cox, J.L. 2014b. 'Can Christianity Take New Forms? Christianity in New Cultural Contexts', in Paul Hedges (ed.), *Controversies in Contemporary Religion: Education, Law, Politics, Society, and Spirituality. Volume 3: Specific Issues and Case Studies*, pp. 195–222. Santa Barbara, CA: Praeger.

Cox, J.L. and A. Possamai. 2016. 'Introduction: The Australian Census, Religious Diversity and the Religious "Nones" among Indigenous Australians', in James L. Cox and Adam Possamai (eds.), *Religion and Non-Religion among Australian Aboriginal Peoples*, pp. 3–23. London and New York: Routledge.

Craig, R.D. 2004. *Handbook of Polynesian Mythology*. Santa Barbara, CA: ABC-CLIO, Inc.

Durkheim, É. 1915. *The Elementary Forms of Religious Life: A Study in Religious Sociology*. London: George Allen and Unwin.

Elkin, A.P. 1964 [1938]. *The Australian Aborigines*. Garden City, NY: Doubleday.

Evans-Pritchard, E.E. 1965. *Theories of Primitive Religion*. London: Oxford University Press.

Freud, S. 1938 [1919]. *Totem and Taboo*. Harmondsworth: Penguin.

Holman, J.P. 2010. *Best of Both Worlds: Elsdon Best and Tutakangahau*. Rosedale, NZ: Penguin.

Irwin, J. 1984. *An Introduction to Maori Religion*. Bedford Park, Australia: Australian Association for the Study of Religions.

Knappert, J. 1995. *Pacific Mythology: An Encyclopedia of Myth and Legend*. London: Diamond Books.

Māori Dictionary of New Zealand. 'atua', http://maoridictionary.co.nz/word/494 (accessed on 19 July 2016).

Marsden, M. 2003. *The Woven Universe: Selected Writings of Rev. Māori Marsden* (edited by Te Ahukaramu Charles Royal). Masterton, NZ: The Estate of Rev. Māori Marsden.

McLintock, A.H. 1966. 'Aotearoa', *An Encyclopedia of New Zealand 1966*, www.tearagovt.nz/en/1966/aotearoa (accessed on 18 July 2016).

Moore, D. 2016. '*Altjira*, Dream and God', in James L. Cox and Adam Possamai (eds.), *Religion and Non-Religion among Australian Aboriginal Peoples*, pp. 85–108. London and New York: Routledge.

O'Dea, T.F. 1957. *The Mormons*. Chicago: University of Chicago Press.

Parsons, M.J. 2010. 'Jury, Hoani Te Whatahoro', *Te Ara: The Encyclopedia of New Zealand*, www.teara.govt.nz/en/biographies/1j6/jury-hoani-te-whatahoro (accessed on 20 July 2016).

Platvoet, J.G. 1992. 'African Traditional Religions in the Religious History of Humankind', in G. ter Haar, A. Moyo and S.J. Nondo (eds.), *African Traditional Religions in Religious Education: A Resource Book with Special Reference to Zimbabwe*, pp. 11–28. Utrecht: Utrecht University.

Reilly, M.P.J. 2004. '*Te tīmatanga mai o ngā atua*. Creation Narratives', in Tānia M. Ka'ai et.al. (eds.), *Ki te Whaiao. An Introduction to Māori Culture and Society*, pp. 1–12. Rosedale, NZ: Pearson Education New Zealand.

Shirres, M.P. 1997. *Te Tangata: The Human Person*. Auckland: Accent Publications.

Smith, J.Z. 1982. *Imagining Religion: From Babylon to Jonestown*. Chicago: University of Chicago Press.

Spencer, B. (ed.). 1896. *Report on the Work of the Horn Scientific Expedition to Central Australia* (4 vols.). Melbourne: Melville, Mullen and Slade.

Spencer, B. and F.J. Gillen. 1897. *The Native Tribes of Central Australia*. Melbourne: Macmillan.

Spencer, B. and F.J. Gillen. 1927. *The Arunta: A Study of a Stone Age People* (2 vols.). London: Macmillan.

Stanner, W.E.H. 2009 [1956]. 'The Dreaming', in *The Dreaming and Other Essays, with an Introduction by Robert Manne*, pp. 57–72. Melbourne: Black Inc., Agenda.

Statistics New Zealand. '2013 Census QuickStats about Culture and Identity', www.stats.govt.nz/Census/2013-census/profile-and-summary-reports/quickstats-culture-identity/religion.aspx (accessed on 18 July 2016).

Statistics New Zealand. 'Census Snapshot: Māori', www.stats.govt.nz/browse_for_stats/people_and_communities/maori/census-snapshot-maori.aspx (accessed on 18 July 2016).

Strehlow, T.G.H. 1947. *Aranda Traditions*. Melbourne: Melbourne University Press.

Strehlow, T.G.H. 1971. *Songs of Central Australia*. Sydney: Angus and Robertson.

Strehlow, T.G.H. 1978. *Central Australian Religion: Personal Monototemism in a Polytotemic Community*. Bedford Park, Australia: Australian Association for the Study of Religions.

Swain, T. 1993. *A Place for Strangers: Towards a History of Australian Aboriginal Being*. Cambridge: Cambridge University Press.

Swain, T. and G. Trompf. 1995. *The Religions of Oceania*. London and New York: Routledge.

Tafjord, B.O. 2013. 'Indigenous Religion(s) as an Analytical Category', *Method and Theory in the Study of Religion*, 25 (3): 221–243.

Te Ara: The Encyclopedia of New Zealand. 'Story: History. Māori Arrival and Settlement', www.teara.govt.nz/en/history/page-1 (Accessed on 18 July 2016).

Te Rangi Hīroa [Sir Peter Buck] 1950. *The Coming of the Maori*. Wellington: Maori Purposes Fund Board.

Te Whatahoro, H. 1913. *The Lore of the Whare-wānanga: Or Teachings of the Maori College on Religion, Cosmogony, and History. Part 1: Te Kauwae-runga, or 'Things Celestial'* (translated by S. Percy Smith). New Plymouth, NZ: Thomas Avery for the Polynesian Society.

Te Whatahoro, H. 1915. *The Lore of the Whare-wānanga: Or Teachings of the Maori College on Religion, Cosmogony, and History. Part II: Te Kauwae-raro, or 'Things Terrestrial'* (translated by S. Percy Smith). New Plymouth, NZ: Thomas Avery for the Polynesian Society.

Trompf, G.W. 1991. *Melanesian Religion*. Cambridge: Cambridge University Press.

Williamson, R.W. 1937. *Religion and Social Organization in Central Polynesia*. Cambridge: Cambridge University Press.

6

INDIGENEITY, THE ENVIRONMENT AND AFRICA

Some key concepts from the /Xam of southern Africa

Michael Wessels

It is impossible, I think, to select a representative set of key indigenous concepts that typify an essentially African mode of environmentalism. One only has to consider the sheer scale of the continent and the extraordinary number of diverse cultures to be found in it. Africa's range of cultures, languages and lifeways is rivalled only by its range of ecosystems – deserts, snow-capped mountains, rainforest, river deltas, savannah – and it rich diversity of flora and fauna. Africa contains a wealth of representations and conceptualizations of the environment; its languages and cognitive patterns were largely formed in its environments, even as it has interacted with the rest of the world since ancient times. Echoing Claude Lévi-Strauss (1964: 89), we might say that the environment – the plants, animals, climate and physical features of the landscape – are used to think with as much as to live in. The numerous ways of thinking with and about the environment in Africa constitutes an invaluable source of knowledge and practice, and also an imaginative resource on which to draw today in a context of the mass destruction of natural environments and the erosion of sustainable ways of living. Even after the ravages of colonialism, rapid urbanization, war and neo-colonial exploitation of resources by the elites in the service of the global corporates, Africa still possesses more wilderness than almost anywhere else, and more people who live lives of relative economic and environmental sustainability within it. Africa's writers, thinkers, artists and storytellers offer a wealth of representations of, and thinking about, what have come to be called the environment.

Where to start? There exists an overwhelming wealth of sources. Numerous anthropological studies exist on Africa's varied cultures, and these inevitably contain information about cosmology and relationships with plants and animals. Increasingly, these are written by insiders – scholars who come from the culture and environment they describe. Africa's rich literatures, both written and oral (and a great deal of African oral literature has been transcribed and recorded, even though this represents only a fraction of the whole), provide another source of information about people's place in the cosmos and about relationships between people, animals, plants and places. These literatures themselves stimulate new ways of imagining indigeneity and the environment, of transcending Cartesian binaries and of spanning the divides between the animal and the human, the living and the dead, and between the cultural and natural worlds. Africa abounds in local literatures that capture particular life ways in specific environments, but also in thinking that looks for a distinctive African sensibility and philosophy across time and place. Then there are the scientific studies of Africa's environments and studies of African knowledge about this environment. Africa has also contributed to environmental thinking and activism through international figures such as Ken Saro-Wiwa and Wangari Maathai.

There is clearly no shortage of material. Two strategies present themselves. One would be the route of generalization. This is the path of the outsiders who have written about the continent as though it were relatively homogenous – this is the case even today in the international press – and also of the Négritude and Pan-Africanist writers and their successors who have identified concepts that seem common to cultures across Africa and which exemplify a common African cosmology and humanism. I shall eschew such a general approach, but not for the usual reasons – that it is essentialist, ahistorical, merely inverts European stereotypes about Africa and tends to downplay Africa's historical openness to the world. These are valid criticisms, but they fail to recognize that this tradition of thinking, as practised by Africans themselves at any rate, has been hugely productive – instilling solidarity, confidence and pride in the face of slavery, colonialism and racism, inspiring new literature and other creative forms, and foregrounding different sorts of knowledge. Most importantly, in my view, a more global view of African culture highlights interconnections that are more than ideological or imagined. People have always moved about the continent and exchanged goods and culture. Despite Africa's varied ecosystems, numerous animals and plant species occur in areas of the continent that are very far apart, providing common economic resources for Africa's people, but also entering the common cultural and spiritual life of the continent.

Nevertheless, my preference here is for the local and specific. I have chosen a group of people who are indisputably indigenous in any sense of the term, the San of southern Africa, specifically the /Xam of the northern Cape in the late 19th century. A people who lost their independent and environmentally sustainable way of life, as well as their language and identity, in the second half of the 19th century. I will draw on concepts from the /Xam language or set of languages and employ these in a metonymic manner to point beyond the /Xam, to other San groups, the contemporary San of the Drakensberg in particular, and also to other cultures in southern Africa. To give an idea of the inexhaustible richness of the environmental location of only one African people will, I hope, point to the richness of the whole.

This essay has to bring together three concepts or terms in order to generate another set of concepts: Africa, environmentalism and indigeneity. Africa provides the geographical and cultural parameters, but it is also, as Valentin Mudimbe (1988) puts it, an invention. It is a geological reality in terms of tectonic plates, but an imagined community when it comes to identity and culture. As I have already mentioned, it is home to countless cultures and languages, as well as to varied environments. These are seldom contiguous with the nation state as is often the case elsewhere. It is a continent of minorities. Africa flows in different directions; it is not always contained within the body of the continent itself. Since the 16th century, it has encompassed the trans-Atlantic world of Africa and the Americas. The continent has been linked to the Middle East and to Asia via Indian Ocean networks for much longer. North Africa is part of the cosmopolitan world of the Mediterranean, and also of the Islamic cultural sphere that stretches east to western China in the north and Indonesia in the south. Africa is part of the world, but the world is also part of Africa. Contemporary world culture is difficult to imagine without African music, for example. The same is increasingly true of written literature. The world was peopled in the distant past by Africans. From the early colonial period until today, very few parts of the world have not been enriched by the presence of modern Africans. African animals and African environments, too, form a vital part of the world imaginary.

The slipperiest of the three terms in the context of Africa is undoubtedly that of the indigenous (see Hodgson 2009). While most African governments claim that all Africans are indigenous, only hunter-gatherers (San and Pygmies) and nomadic pastoralists (Tuareg and Masai) have been officially recognized as indigenous by world bodies. Adam Kuper's (2003) argument that the use of the term "indigenous people" signals the return of the primitive and leads to a divisive identity politics based on notions of cultural purity has been countered by arguments that the "strategic essentialism" (Spivak 2006: 205) that attends the politics of indigeneity is indispensable in struggles for economic and political autonomy on the continent. Worldwide the term is contested, but nowhere more so than in Africa and Asia, with their complex precolonial histories of migration and mixing. The term is often used to mean the original people; the people who were there first. Of course, almost all humans moved from somewhere else and we cannot know of all the displacements across history. But indigenous people got there before the others or, as is the case with the Khoi of southern Africa, are related to those who did. The San, possibly alone of all the world's people, appear to have been always here. We should not be too quick, though, to conflate the San encountered by Europeans with the prehistoric peoples of southern Africa, as though they were immune to history and to change. The San are very much a people in history. Archaeological, linguistic and genetic records reveal fluctuating settlement patterns, migrations, trade and interrelationships with other groups over thousands of years.

The term "indigenous" in Africa is relative: relative to European settlers, Africans and African languages are indigenous; relative to mainstream cultural and linguistic groups, people like the San or the Baka and Bagyeli are indigenous. Fortunately, the question as to who is entitled to the designation of "indigenous" is not critical to the environmental concerns of this chapter, however. All the cultures of Africa have their roots in precolonial lifeways that were, for the most part, environmentally sustainable and possess concepts and ways of thinking about the world and the environment that offer a wealth of practical and philosophical resources for today. From the beginning of human life on Earth until the advent of European colonialism, people in Africa coexisted with almost inconceivable numbers of wild animals in a great variety of habitats. None of the ways of life pursued on the continent appear to have extensively undermined the integrity of the environment. Even the precolonial trade to the east in ivory and rhino horn did not significantly change that. There is the argument, too, that anthropogenic burning actually increased the quantities of wildlife by extending and maintaining the animal-rich savannah (see, for example, Laris et al. 2015). Even today, many Africans pursue a subsistence form of agriculture, in conjunction with various forms of foraging from the wild, that is relatively sustainable. So why have I chosen the San for this chapter and why the /Xam, in particular?

The San are the quintessential indigenous hunter-gatherers, famously able to subsist in hostile environments. They have long been recognized as close to nature, a status that once justified genocide and land dispossession – they were represented as closer to animals than to humans and apparently had no conception of property. Their "natural" way of life, though, has also proved a source of endless fascination to outsiders and there exists today a vast body of anthropological literature on the San on which to draw, most of which is based on communities that pursue a hunter-gathering lifestyle or that still recall the practices and culture that attended it. If I were an anthropologist, I might have turned to living San informants to learn how they themselves might describe the ways in which their thinking is informed by categories drawn from the natural environment. Instead, I chiefly rely on the San oral literature and history that was recorded and translated in the second half of the 19th century by the linguist Wilhelm Bleek and his sister-in-law Lucy Lloyd from /Xam informants who had been imprisoned in Cape Town. The very rich /Xam archive, possibly the most extensive body of recorded oral literature in the world in a language that no longer exists, forms the core of the essay. This choice is not unproblematic. /Xam literature was collected in the ideological climate of Social Darwinism at a time of genocidal attacks on the San by commandos and the capture and enslavement of indentured San children by farmers. The term "San" has mixed connotations, some of them negative, as does the interchangeable term, "Bushman." People identified themselves as belonging to smaller place-based groups such as the /Xam or Ju'hoan rather than to a generic ethnic group, the San or Bushmen. People became San, and later indigenous, through a history of loss and cultural, linguistic and economic dispossession. These were identities conferred on groups of people by outsiders, although they were later embraced by San people themselves as part of a strategy of solidarity and resistance.

The idea of the environment is a modern one – the product of a certain separation from life lived on the land and the formations of consciousness that attended the shift to urbanization. Environmentalism is rooted in an awareness and politics of loss and conservation. /Xam thought, as it is reflected in the primarily precolonial world of the Bleek-Lloyd archive, did not possess a set of self-reflexive environmental principles. Nevertheless, concepts can be gleaned from /Xam thought, literature and practice that have value for us today as environmental principles. The /Xam were not proto-conservationists. The San, in general, were not above participating as actors in the networks that were involved in the decimation of the wildlife of the region. Robert Gordon (1992) has shown how the nearby San in Namibia participated in the 19th century ivory trade. Nevertheless, the /Xam were hunters and gatherers who inhabited a social and physical space that was the same space inhabited by wild animals and plants. Their culture – rock art, narrative, music, religion – cannot be separated from the environment. Their economy was dependent on the rich biodiversity of a wild environment and was also ecologically sustainable in every way. The /Xam lived in small bands near permanent waterholes in an arid land, supplementing their diet of bulbs and other food plants, collected chiefly by the women, with animals hunted chiefly by men. Such a lifeway is environment-specific, reliant on the intimate knowledge of the edible plants of a particular habitat and the movement of wild animals within it. It relies on the regeneration of the plants and animals in the environment. The /Xam, for example, never took all the eggs from an ostrich nest or all the honey from a beehive. As we will see, the practice of sustainability was accompanied by an ethos of interconnectedness in which the human and animal, the living and the dead, and the natural and the social were united in a web of relationships, mostly benign but sometimes malign.

Today, more and more people are beginning to claim a broader Khoisan identity in South Africa. Khoisan identity encompasses the legacy of all the precolonial inhabitants of the winter rainfall areas of southern Africa as well as that of San groups further north and mostly pastoralist Khoi groups that migrated out of the Cape. The Khoisan heritage of large numbers of people is becoming increasingly evident as links are shown to exist between contemporary oral literature in Afrikaans, for example, and that of the San of the 19th century (De Prada-Samper 2016). The "secret San" of the Drakensberg now claim a San identity after decades of concealing it in order to avoid discrimination. They retain a sense of San ancestry, but speak the languages of the Sotho, Xhosa and Zulu communities into which they have been assimilated (Francis 2007; Prins 2009). Unsurprisingly, the politics of Khoisan identity is fraught with tension. Authenticity is contested. Most people in the region are part San or Khoi, but few are completely so. Only small groups in Botswana and Namibia pursue a foraging way of life, although larger numbers of people in those countries speak a San or Khoi language.

In this context, one has to ask the question about the validity of an outsider raiding San culture for concepts that can be made to speak of an indigenous environmentalism. The concepts I have chosen are unlikely to coincide with those that a /Xam thinker would have identified. My selection is small and to some extent arbitrary, determined by the ways in which the /Xam archive has been read and my own familiarity with it. How does one evade nostalgia for a lost innocence when dealing with such materials? To what extent is what Antjie Krog (2009) calls the San interconnected worldview a product of Western longing? These are important questions which I have dealt with elsewhere (see Wessels 2010). Here, I will concentrate instead on the fact that /Xam literature exemplifies an experience of the world in which people belong to a place and an environment rather than the other way round. San knowledge is embedded in very specific living environments, inhabited by particular species of plants and animals. San stories refer not to generic frogs, for example, but to specific species of frogs. Large and small animals enjoy an equal place. The boundaries between the human and the animal are fluid. Nature and society intersect. Nor is it always easy to distinguish between the abstract and the concrete. Dorothea Bleek (1929: 9), Wilhelm Bleek's daughter and intellectual heir, observes that: "The whole animal world is very much alive to Bushmen, the border line between the powers of nature and animals is vague, that between animals and man more so." Ideas about the hard division between culture and nature are alien to /Xam thought. The stories are full of crossings between the human and the animal. In the next section, I will explore this phenomenon in more detail and briefly discuss its implications for thinking about the environment today.

Animals

Animals of all shapes, sizes and metaphysical standing form the imaginative, aesthetic and ontological core of most San paintings, engravings and stories. Many San stories only involve animal characters. The stories of the early times involve characters which are at once human and animal. People not only eat and wear parts of animals, but hunting, dance and storytelling all involve close identification, even merging, with animals. Human life, thought and culture for the San would be inconceivable without animals. Animals are concepts and signifiers as much as they are flesh-and-blood biological organisms. Drawing on ideas of "new animisms," Mark McGranaghan (2014a: 674) notes that the way communities like the /Xam "commonly assign 'personhood' status to a range of non-human groups" enables them to make "[e]valuations of appropriate and inappropriate forms of behaviour . . . [that] incorporate interactions not only with other (human) people, but also shape encounters with non-human species." This forms the cornerstone of an ethos based on a human-animal continuum in which "physical descriptions and personal traits" are "linked to positive and negative assessments of personhood" (ibid., 673).

The San were hunters who especially prized the flesh of large antelope such as the eland. Nevertheless, it was not meat but plants on which they most depended for food. In the Kalahari today, San people still eat more than 80 species of plants.

It is physically possible that the San of areas rich in plant life, such as the Drakensberg, might have subsisted without hunting. But San culture, rock art and narrative is unthinkable without hunting because hunting is not only an economic strategy, but a means of conceptually mediating a great many life phenomena, including gender and sexuality, illness and healing, and relations between people and animals. Unsurprisingly, hunting incidents form the basis of many stories. More importantly, though, the structure of the hunt is replicated in the structure of San narrative itself. People leave the domestic space for the hunting grounds where they find not only animals but stories. The resolution of the story frequently involves a return home.

Animals in San literature do not merely serve as metaphorical substitutes for human attributes and relationships. This is highly significant, in my view, because it is indicative of a more ecological view of the interconnectedness of the animal and human worlds than literatures in which animals stand in for humans in order to exemplify or exaggerate human characteristics. Animals, as represented in San stories, are beings in their own right, with consciousness and judgement. They engage with human beings but are also different from them. The divide between human and animal is continually put under pressure. /Xam storytellers, and San storytellers today in the Kalahari and Namibia, become in part the animals of which they tell. Metamorphosis and other sorts of interplay between human and animal are central to the protean signifying capacity of the stories. Artists do not only paint animals, but use animal ingredients in their paint, some of these for magical reasons. The boundary between the human and the animal is not only crossed in art and literature, but forms part of lived experience. In trance in the Kalahari, San people harness the energy and power of animals, a force that is both spiritual and physical. The hunter's arrow bridges the distance between human and animal in a way that is as much metaphysical as it is physical. After an eland has been struck by a poison arrow, a sympathetic relationship is established between the animal and the hunter. In death, humans, after a time as stars, become fused with animals in an underwater realm from where they continue to interact with the living in ways that can be malign or benign. They can help heal, hunt and make rain, but can also harm and kill. Clearly the differences and similarities between human and animal are aesthetically, epistemologically and metaphysically enormously generative.

In /Xam stories, animals speak to humans, trick them, fight with them, help them and turn into them. The world of San stories is a world in which the boundary between animal and human is fluid; it is constantly subject to revision and negotiation. /Xam stories are stories of transition and becoming. They tell of a process of greater differentiation and separation as the order of the first time gives way to the order of the present. But they are also reminders of the fluid nature of being in the present.

Many species of animal double as characters and as concepts in San literature, just as they double as animals in the environment and as concepts in everyday life. I shall now go on to discuss three of the most important – eland, lions and baboons – providing some pointers as to how these species work conceptually in /Xam thought and practice.

Eland

Eland, the largest antelope species on Earth, occur in all the environments inhabited by the San; they are as powerful a presence in the arid semi-desert of the northern Cape as they are in the lush mountain valleys of the Drakensberg-Maloti. They are the most prized prey of the San hunter. Qing, a 19th century Drakensberg San informant, compares them to cattle (Orpen 1874: 2–3). And, as is the case with cattle for the agro-pastoralist of the region, eland are much more than flesh-and-blood animals. Their central place in the San imagination is reflected in their prominent place in narrative and in rock art. There are more depictions of eland in the rock art of the Drakensberg than of any other animal. They are a common motif in the rock engravings of the /Xam area, too. Eland are good to eat, but clearly, they are even better to think with. It is not easy to fix the meaning of "eland," however; it is an extraordinarily protean signifier. Eland have been linked to sexuality and gender (Biesele 1993), to shamanic transformation (Lewis-Williams 1981), to San social formation (Vinnicombe 2009), and to the inauguration of a world divided between the living and the dead (Solomon 2007). David Lewis-Williams (2015: 83) summarizes the "symbolic associations" carried by the sign eland among the Ju/'hoansi in Lorna Marshall's work of the 1950s as follows: "meat, fat, health, strength, rain, fertility, plenty, unaggressive behaviour, and general well-being – all life-giving things."

The creation of the eland is a central narrative in different San traditions. Both the /Xam and Drakensberg-Maloti accounts feature /Kaggen, the Mantis, a character central to creation stories of the people of the early race. In the /Xam versions, of which there are several, /Kaggen steals his son-in-law's shoe and fashions the eland from that. He hides the baby eland in the reeds next to a waterhole and rears it on a diet of honey. Lewis-Williams (1997: 204) argues that the "eland will become the shaman's power-animal *par excellence*." In this reading, the story concerns the origin of shamanism: "in making this mediatory, potent animal, the Mantis created the basis for /Xam shamanism" (Lewis-Williams 1997: 205). In the course of the story, the Mantis also makes the moon, again by throwing a shoe into the sky. The moon also works as a key element in /Xam thought, and is especially associated with death and regeneration.

Eland, in terms of Lewis-Williams' trance hypothesis, conceptually represent spiritual transformation. In trance, people lose themselves and submit to a power beyond themselves. Dying and trance are frequently associated. In trance, people become more than themselves – a state signified by merging with the animal, frequently represented in rock art by the hybrid form of the therianthrope. Not all scholars interpret these images as depictions of the transformations undergone in trance, though. Anne Solomon (2007), for example, reads them as depictions of the spirits of the dead. Once again, though, the transformation of a person into a spiritual being is represented by their fusion with an animal, most frequently the eland.

The eland is also conceptually linked to the ambiguity of hunting: people kill and eat that which they most revere. Eland feature centrally in a complex of everyday beliefs and practices that involve /Kaggen in his role as guardian of the animals. He tries to prevent a group of animals, the most prized meat animals, from being killed. Chief among these is the eland; the hartebeest, the gemsbok, the now extinct quagga and the springbok are the others. /Kaggen, it is said, does not love people when they hunt these animals. Rituals have to be performed to ensure success in the hunt in the face of /Kaggen's opposition (Hewitt 1986: 125). He might, for instance, assume the form of a snake in order to startle hunters and alert the prey animals to their presence. His most telling interventions occur after an eland has been struck by a poisoned arrow and before it dies. During this time, a sympathetic bond is established between the hunter and the animal so that the condition of the hunter parallels that of the animal. /Kaggen exploits this relationship in order to enable the animal to fight off the effects of the poison. He might, for example, startle the hunter by becoming a puff adder, so infusing the wounded eland with more energy (Hewitt 1986: 127). The man who has shot an eland has to return home as if he himself were the wounded animal. The body of the hunter serves as a conduit through which the forces that weaken or strengthen the animal pass. As Hewitt (1986: 126–127) notes "the emphasis was always on the hunter as a medium through which messages might be transmitted to the game." Eland in this context conceptually signify the interconnection between elements in an ecology involving exchange and reciprocity. The hunter is fused, for a time, with the life he is taking.

There is also a conceptual link between eland and sexuality, and eland and gender roles. Parkington (1996: 282) emphasizes the sexual significance of eland: "The eland has a quite specific connotation related to the parallels seen by hunter-gatherer people between hunting and sex, and the roles these activities play in defining social roles." Megan Biesele (1993) explores this phenomenon among the Ju'hoan of Botswana in her fine book *Women Like Meat*. Women are pursued by men, but are also highly valued – as meat is. The male role of going out, hunting and bringing home meat is equated with the male sexual role. Nevertheless, placing women in the position of prey and men as hunters does not refer so much to gender-determined power relations as it does to the differentiation required for the maintenance of proper human relations. The symbolic female equivalent to male hunting, observes Biesele (1993: 137), is "menarche and marriage." Both male and female activities are central to the reproduction of the social order: "Men's activities are shown to be complementary and indispensable to those of women." Biesele describes a complex negotiation through storytelling of sexual power between men and women. Biesele's title also refers to women's expectations of men. Men who fail to bring home meat for women invite scorn and rejection. Women police the performance of masculinity and continually evaluate men's economic performance. They are not helpless prey.

The signifier "eland" does not function on its own conceptually. In relation to gender and sexuality, for example, it works in conjunction with honey and fat, also concepts as much as they are prized material substances. Honey, notes Sigrid Schmidt (2013: 74), is symbolically central not only in San life but "in African lore and life" more generally, standing for "life force," and eating honey or fat is analogous to having sex (Biesele 1993: 86).

Lions

Given the symbolic resonance of the lion, even in parts of the world where they do not occur, it is unsurprising that this animal plays a vital conceptual role in San thought. Even a quick survey of the references to lions in the /Xam materials reveals that it is a protean sign that occurs in a variety of contexts. Some of the references clearly have a supernatural import, and many involve transformations that could be linked to shamanism; Lewis-Williams and Challis (2011: 138) claim that "any lion encountered in the veld could be a transformed shaman." Mathias Guenther (1999: 187) maintains in the case of the Kalahari that "the lion was the trance dancer's most common incarnation." In a number of cases, argues Anne Solomon, spirits of the dead take the form of lions rather than live shamans. In general, these spirits are intent on harming humans. The lion, observe Lewis-Williams and Challis (2011: 135), "was much more than a dangerous predator. It was fearfully associated with the spirit realm." Some of this can be attributed to its nocturnal nature. It was believed that lions "had the supernatural ability to cause the sun to set and thus to bring darkness in which it could conceal itself" (Lewis-Williams and Challis 136). Unlike the supernatural connotations of the eland, those connected to the lion are often threatening. Patricia Vinnicombe (2009: 210) sums up the basic conceptual divide the /Xam established between large feline predators and antelope:

> lions and leopards were associated with harm as opposed to benefit, with disease as opposed to health, insecurity not security, malevolence rather than benevolence, with death as opposed to life . . . Antelope were regarded as a constructive force in Bushman symbolism. Lions and leopards were destructive.

While spirits of the dead or shamans might assume the form of a lion, lions, in turn, have the ability to turn into other animals and even people. One of the most common reasons for this is when the lion wishes to trick a person and then attack or eat them. /Xam informant |Han#kass'o describes how a lion turns into a hartebeest in order to trick people into hunting it. The hunters become the prey when the animal reverts to its true form (L.VIII.23:8075–77;[1] Hollmann 2004: 62). An attack on a girl is described in which a lion tries to trick the girl into pulling him up onto the rock on which she is hiding by speaking like a person (Hollmann 2004: 37–40). People can turn into lions and lions into people or other animals. But lion transformation doesn't stop there. Different animals can take the form of lions. In one incident, wildcats that had been killed by hunters in the daytime became lions that night and attacked the hunters (L.VIII.23:8080–83, 8082'; Hollmann 2004: 60). In another account (L.VIII.31:8775–82), a gemsbok that is being stalked by a man turns into a lion.

Many of the transformations involving lions represent danger. Conceptually this is consistent with the real threat lions represent. Lions not only provide competition for game, but are also an ever-present danger to human life. Lions, according to ||Kabbo (L.II.10:1053–56), will kill and eat all animals, humans included. Humans have to be vigilant and constantly interpret the signs that might indicate danger. Certain behaviour by owls and crows, for instance, represents forewarnings of danger from a lion (Hollmann 2004: 41–44).

Apart from the steps that need to be taken when a lion is planning an attack, there are certain precautions that always have to be taken in order to prevent a lion attack in the first place. A torn water bag might be the work of an invisible lion that will turn into a tangible one and lie in wait for his victims at the waterhole (53–55). People, especially children, should avoid using the word "lion" in case the flies hear what they are saying about the lion and report it to him. He would then be angry and attack at night. People should never take the whole of a lion kill if they come across one, but leave the head and the backbone. The meat they find should not be cut up in view of the lion. A lion that was left with no part of its kill would consider that it had been treated without consideration and track and harm the men who had taken all its meat.

Guenther (1999: 71) attributes this behaviour not only to fear but to a relationship of mutual respect between lion and human. Given their similar niche in the ecology, a negotiated coexistence is the only way to prevent outright conflict and continual competition. Lions and humans hunt the same antelope species and live near the same waterholes. Lions share the basic ambiguity of all animals; they are like humans but also different from them. Lions think like people; they also share a number of habits with people. Both people and lions are predators capable of killing big game such as eland and gemsbok. Neither predator consumes its prey at the place it kills it (Hewitt 1986: 47). According to the most senior /Xam informant, //Kabbo, lions "talk, they are also people" but they are "people who are different" (quoted in Lewis Williams and Challis 139).

Conceptually, similarity and difference are vital, as McGranaghan (2014b) has argued. Unlike the eland, which represents "all life-giving things," the connotations of the lion are chiefly negative. They are like people when people do not behave well; they are angry people. This suggests that violence and anger separates people from the human and transports them into the realm of lion identity. Lions in this way enable thinking about the sort of violent human behaviour that is driven by passion rather than thought, such as in the case of male domestic violence. Such behaviour, in both a person and a lion, results from a "lack of understanding" (10). The /Xam believed that people who were "'possessed of their thinking-strings' (i.e. acting in accordance with the norms of a rational, socialized person) would not engage in violent behaviours." Thinking about lions also helped the /Xam to distinguish themselves from groups that attacked them, such as the European colonists and the neighbouring Koranna pastoralists. The seemingly unmotivated violence of the European settlers towards the San aligns them with lions (13).

McGranaghan (2014b: 5) notes that "hunter-gatherer groups" are often "characterized as possessing ontologies that extend 'personhood' to a range of non-humans." While this recognizes the fluidity of identity between people and animals, and the possibility of transformation, it also allows the conceptualization of positive and negative forms of human behaviour. A human who engages in negative behaviour takes on certain undesirable animal traits. In this case, those of lions, becoming in the process something of a monster. Predators, in particular, provide "powerful idioms for antisocial, monstrous behaviour" (10).

Baboons

In Dorothea Bleek's view, there "is no great divide between man and animal in their [San] thinking" (Weintroub 2016: 149). Nowhere is the divide between human and animal as thin as it is in the case of baboons. The San recognized that baboons and people share a great many characteristics. Dorothea Bleek notes that this realization would mean that the idea of descent from apes, still a shocking idea to many Europeans at the time she wrote, would "not be unfamiliar or repulsive" to them (162).

The chacma baboon, a species common throughout the area inhabited by the San in southern Africa, features prominently in stories. Conceptually, baboons explore the play of difference and similarity between people and animals in both the /Xam First Times narratives and the discourses of the present order that appear in the Bleek-Lloyd materials (Hewitt 1986: 38, 109–110; Guenther 1999: 74; Hollmann 2004: 7–9). Guenther (1999: 74) lists the characteristics baboons are said to share with people: "[b]aboons have wives like humans; they have speech, as well as songs; and they understand the human language and call the Bushmen by their names." This produces a high level of "moral ambiguity" when it comes to killing baboons, which, it should be noted, are hunted for medicine and not for food. The ambivalence which attends the hunting of baboons means that a great many precautions are followed in hunting them. Parts of the baboon – its teeth and eye hollow, in particular – would remain in the hunter's bow unless he cut lines around its point before hunting them. Even more seriously, the "baboon's death would live in our bows" (20). Because baboons were once men, they react like men when they are being hunted (21). A baboon can also shoot an arrow back at the hunter who has shot it. To prevent this, the hunter should tell the baboon that it is a girl's arrow with which it has been shot. This makes it ashamed, and it accepts its death (22–23).

Baboons are said to have been people even before people existed (Hollmann 2004: 24). "My parents used to say to me, that the baboons were once people at the time when we who are people were not here," reports the /Xam informant Dia!kwain. Baboons are not alone in once being people. What makes them different, though, is that they retain a consciousness of this history and also retain several human physical features: "their parts resemble humans, for they feel that they are people. That is why their parts smell of people" (Hollmann: 24). People "make clouds when they die" (21) and so does a baboon because "it feels that it too was once a man" (20).

One of the distinguishing characteristics of humans is their love of play and games. This requires the separation of a time for play from that of ordinary life. In San stories and beliefs, baboons share this trait. In fact, it is baboons who taught people one of their most popular games, the #gebbi-gu – a "game or dance with peculiar calls" (26). The baboons still play this game (Hollmann 2004: 24–28) and come to watch people play it. One of the best known of the /Xam stories features baboons playing a game. The baboons punch /Kaggen's son after he admits that he is looking for a stick which his father will use to threaten them (Bleek and Lloyd 1911: 21–23). The boy's eyeball falls out when he is punched and the baboons then begin to play a game of ball with the eyeball, singing a playful song as they do so. /Kaggen manages to retrieve the eyeball after he joins in the game himself. When /Kaggen finally returns home and recounts the story, his grandson the ichneumon scolds /Kaggen for mixing with strangers. The point is that a relationship of respect and distance should be obtained between the human band and the baboon troop because baboons are people who are different. A lot of the information about baboons in the /Xam materials reinforce this point. Baboons can speak /Xam, but one should not speak to them (Hollmann 2004: 10–16). This requires awareness on the part of humans because baboons desire to talk to them: "it is thing which does not merely want to see us, to look and leave us; but if it catches sight of us, it talks to us as long as it sees us, for it wants us to talk to it" (Hollmann 2004: 16). Special care should be taken not to mistake a group of baboons for humans when one approaches from a distance. One should not talk negatively about baboons because they might hear and take offence.

Hollmann (2004: 7) suggests that the need for people to separate themselves from baboons might have had "an ecological component," since the two species often subsisted on the same plants. Just as lions compete for prey with San men, baboons gather plants as San women do. The competition for plant resources is only part of the story, though, because plants are not only food but medicine. Sam Challis (2012) notes how baboons in the Eastern Cape in the 19th century were seen to have special powers because of their intimate knowledge of the powers of plants by the AmaTola, an ethnically mixed group of stock raiders with a strong San component. Certain roots in particular were believed to offer baboons protection from harm and also "enabled baboons to sense approaching danger" (Challis 2012: 276). The rock art of the AmaTola indicates, in his view, that people could harness this power; in it there are scenes in which "dancers enter trance to acquire baboon potency and mediate with the spirit world" (278). The /Xam informant Dia!kwain describes how a baboon will hold a stick from a plant in its mouth that "tells it about things which it does not know" (Hollmann: 10). These sticks warn them of danger and also protect them from pain and illness. Not surprisingly, the Bushmen try to get a hold of these sticks from the baboons (Hollmann: 11).

While baboons share characteristics with humans and possess powers that humans desire, they also serve to identify categories of humans one should avoid, especially in relation to marriage. Hewitt (1986: 109) claims that baboons "represent the stereotype of undesirable in-laws. They are seen repeatedly as undesirable neighbours whom one would certainly not want one's daughter to marry." In a story from the Drakensberg, a baboon wishes to take a young girl as his wife (Orpen 1874: 6–8). The girl takes refuge with a reclusive chief named Qwanciqutshaa. The baboon fights Qwanciqutshaa for the girl, but is defeated:

> Qwanciqutshaa got it down and stuck it through with his own keerie, and Qwanciqutshaa banished it to the mountains, saying, "Go, eat scorpions and roots as a baboon should," and it went screaming away; and the screams were heard by the women at the place it came from, and all the baboons were banished.
>
> *(Orpen 1874: 7)*

The problem in this story is not the animalism of baboons, but the opposite – their closeness to human beings. While the marriage between the baboon and the human girl fails to materialize, it is, for a time, a possibility. Indeed, Qwanciqutshaa describes the baboon as the girl's husband who comes from her "place," unlike Qwanciqutshaa himself, who lives in another place. The baboon's different identity becomes fixed in terms of its different physique and its diet – its culture, we might say, rather than its animalism.

From this short overview it can be seen how the baboon functions as a key concept in San life. For a start, they help people to think about that which can be hunted and which cannot, the phenomenon of the stranger, marriage and the supernatural power of plants.

/Kaggen, the Mantis trickster of the /Xam

/Kaggen, or the Mantis, sometimes described as a trickster deity (Lewis-Williams 2000: 8), is the central character in many /Xam tales of the First Times, a fictive or formative period in which animals had not yet been separated into their species or separated from people. The characters are animals who can speak and live like humans but who also display some of the characteristics of the animals they will become. /Kaggen also appears as a figure who protects certain large animals he especially values in everyday life. In many of the /Kaggen stories, Hewitt (1986) notes, the violation of social rules occurs away from home. While this structure supplies the "formal principles" for the generation of the narratives (181), /Kaggen himself, is the "operational principle" which enables the fundamental theme of the relationship between the social order and its disruption to take shape.

/Kaggen is a signifier which works conceptually in multiple ways. /Kaggen is indicative of a way of life characterized by fluidity, adaptability and uncertainty, argues Guenther (1999). He exemplifies an ambiguous, open-ended worldview suited to a foraging economy. /Kaggen represents the multifaceted quality of the natural world. He is simultaneously "destructive and creative" and embodies a cosmology that is a blend of "the numinous and the ludicrous" (227). He is not an "architect of order and structure" (121) like the grim, unsmiling God of the middle-eastern monotheisms, but a "figure who embodies self-contradiction, a spirit of disorder and who is an enemy of boundaries." Guenther relates this to the environment, which is both a source of life and of death. Animals are a food source, but they also represent danger – both physical and spiritual. The environment is unpredictable and the forager needs the flexibility and ingenuity of the trickster to function within it. Individual opportunism is often necessary, rather than the sort of social conformism more rigidly structured societies, with their law and order deities, encourage. But /Kaggen tests the limits of individualism as well. The environment of the forager demands both individual ingenuity and a commitment to the collective in terms of the equal distribution of resources and a readiness to cooperate in procuring food.

In the stories of the First Times, /Kaggen lives with his family in a world shared by people who display "some of the attributes of the animals they will in time become" (Hewitt 1986: 105). The narratives tell of the separation of the species, the loss of human characteristics and the confirmation of the attributes most associated with a particular animal. In doing so, they depict the difference between species and also between animals and humans while simultaneously emphasizing the fundamental similarities between them. All species are related by virtue of their common ancestry in the period of the First Times. Many of them still possess features from this time. Part of the body of the hare, for example, is still human.

The stories of the First Times are underpinned by an ecological and democratic vision that also informs life in the present. The species are interdependent and interact on physical and spiritual planes. Various kinds of transformation from one species to another are common. There isn't a hierarchy of animals. The long-nosed frog, for example, is as important in the stories as a lion or an elephant. A wide range of species (nearly 50) are represented in the narratives, including insects, birds and mammals of every variety. All occur in the /Xam area and represent real species rather than types or analogues for human characteristics. Nowhere is the egalitarian and heterogeneous nature of the animal world as evident as it is in the wonderful mix of species that comprise /Kaggen's family. The Dassie is his devoted, but critical, wife. Their adopted daughter, Porcupine; her husband, Kwammang-a (associated with the rainbow, in which he can be seen); their son, the Mantis's grandson, Ichneumon (a type of mongoose); and /Kaggen's sister, the Blue Crane, make up the immediate domestic circle.

In many of the stories, /Kaggen encounters another group of animal-people and ends up fighting with them, often provoking their hostility through his tricks or his failure to observe protocols governing behaviour with strangers. He usually comes out the worst and has to escape from danger by resorting to magical flight. He is the only /Xam figure that truly possesses magical powers, including flight, transformation, creation and resurrection. He is usually admonished by members of his family when he returns home for making trouble with other beings. /Kaggen, though, is a complex character. In some of the stories, he plays a protective rather than a trickster role.

Besides being a central character in the /Xam stories of the First Times, / Kaggen also plays an important part in /Xam practices and beliefs concerning hunting. He tries to interfere in hunting of eland, hartebeest, gemsbok, quagga (an extinct species of zebra which was once very numerous in the Cape) and springbok. He is said to not love people when they kill these animals. His most telling interventions occur after an eland, a hartebeest or gemsbok has been shot, in the period between the time the arrow strikes and the animal's death from the effects of the arrow poison. During this time, a sympathetic bond is established between the hunter and the animal. The condition of the hunter parallels that of the dying animal. /Kaggen exploits this relationship in order to enable the animal to fight off the effects of the poison and recover.

As this short summary illustrates, the figure of /Kaggen covers a great deal of conceptual ground. He enables, for example, thinking about the human relationship with the natural order. His contradictory character reflects the complex nature of reality. He is a creator and possesses powerful magical abilities, and yet his adventures frequently end in disaster – a warning against hubris. As Hewitt argues, he serves as an example of how not to behave, especially when interacting with others in the world outside the family group, which in /Xam life is virtually synonymous with what we would call the environment today – a world of different but interacting animals and plant species, and natural features such as water holes and hills. The admonitory lecture delivered to /Kaggen at the end of his disastrous adventures also serves to reinforce the social order for those listening to the stories. At the same time, the hilarity that attends his doings undermines an overly serious view of life and celebrates the unpredictability of existence. His role in hunting, on the other hand, reinforces codes of respect for the animals that are being hunted and the procedures surrounding hunting. It signifies that life is a set of relationships in an environment. /Kaggen's role as protector of the game helps conceptually mediate anxieties about killing animals. Animals, after all, possess spiritual and other sorts of potency. /Kaggen embodies both the best human attributes – protectiveness, fearlessness, ingenuity, playfulness – and, more frequently, less noble qualities such as lust, violence, boastfulness and greed, the last a particularly reprehensible quality in a harsh environment in which survival depends on an ethos of sharing.

!Khwa

!Khwa is a /Xam supernatural being who represents the unpredictable and contradictory qualities of nature. He is especially associated with the destructive force of "male" rain, which can be contrasted with female rain, the nurturing power of gentle rain. !Khwa, according to Anne Solomon, often appears as a "deeply sinister figure" in /Xam narratives and can be equated with "death itself" (Solomon 2007: 46). Jeremy Hollmann (2004: 131) interprets him as a more ambiguous figure who is not altogether unrelenting in his opposition to humans:

> !Khwa was a conscious, gendered being who controlled where and how the rain fell. His behaviour was largely determined by people's behaviour – when they respected him, he favoured them with 'female' rain and harmonious social relationships; when people (especially young unmarried women) did not respect him, !Khwa became angry and destructive. Every member of the community had their own relationship with !Khwa: which, in turn, affected everybody else. Most important for men was their ability to bring female rain to where it was needed. Women seemed to have a physiologically-based relationship with !Khwa, focused upon menstruation, which gave them !Khwa's magic power.

!Khwa, according to Hewitt, "operated exclusively at a point of transition within society," specifically the rites of passage of a new maiden relating to the onset of menstruation (Hewitt 1986: 79). He was linked closely to water things, cobras and puff adders, as well as to turtles and tortoises. If an unmarried women ate a tortoise, for example, the thunder would come searching for her (L.VIII.26. 8304–8309).

In several stories, the wrath of !Khwa, the rain "god," is provoked by a girl's violating the seclusion rules that relate to first menstruation. Between menarche and marriage, unmarried girls should avoid the rain even when they are not menstruating (Hewitt 1986: 282). The rain smells a woman and comes to her as mist in one of the stories. She is overwhelmed by the sweetness of his fragrance. He abducts her, but she manages to escape after putting him to sleep by rubbing his forehead with buchu (L.VIII.16. 7434–7448). Dia!kwain adds in a note that the young woman's "understanding was that with which she worked the Rain nicely; and this was why all the people lived, who would have been all dead; all would have been frogs" (7448). A direct refusal to comply with Rain's wish to take her would have resulted in the gravest consequences – the death of the community in human form and their transmogrification into frogs. !Khwa helps conceptualize the ambiguous power of female sexuality and fertility, especially at the liminal time between girlhood and womanhood, and also the intersection of the human with powerful natural forces.

The /Xam were not entirely at the mercy of !Khwa. It was possible to harness the powers of nature through intercession with the ancestors. Rainmakers could influence the rain to fall in a certain place where it was needed. There are several accounts of the rain-bull, "a creature or manifestation of !Khwa" (James 2001: 225), being led across the countryside and cut as it walked. Rain would fall on the places where the creature's blood touched the earth. There is some debate as to whether rainmakers were living people or spirits of the dead with whom the living could intercede. Whatever the truth, rainmaking works in the interstices of the social and natural worlds, signifying the interpenetration of human knowledge, supernatural power and natural forces.

!Khwa is a reminder for us today of the ambivalent power of the environment – of its power to both sustain and destroy. Nothing in the parched world of the /Xam was more important than water and yet nothing was as dangerous. The danger of a violent storm, with its lightning and flash floods, is obvious, but !Khwa was also associated with the danger of the more benign-seeming waterhole. Some of this danger probably originates in the fact that the waterhole is a place in which different species come together. The waterhole, though exemplifying the force of life, was also associated with the dead. The world of the living dead lies below its surface. It is an intermediary place and a site, therefore, of unpredictable transformations.

Tapping

An intimate relationship exists between a person's body and the environment. The internal system of the body is directly interconnected with the external world of nature beyond the body. //Kabbo, the oldest of the /Xam informants, terms the information that people receive through their bodies "Bushman letters." Tapping is the Bushman equivalent, he contends, of books or letters, since "tappings" in the body "take a message" or convey an account of what happens in another place (Bleek and Lloyd 1911: 331). Bushman letters "speak, they move, they make their (the Bushmen's) bodies move" (Bleek and Lloyd 1911: 331).

A great variety of information is gained from tappings. A man might know, for example, that his father is approaching, for he feels a tapping in a place on his body that corresponds to a wound on the older man's body. A man knows he will kill a springbok when he feels a sensation in his calves that corresponds to the feeling of blood dripping down his legs when he carries a slain springbok. A sensation is felt in the feet when the springbok come, and another in the head when hunters are busy cutting off springbok horns.

Tapping works conceptually to make apparent and decode the connections between things – the web of ties that link people with other people and with animals and the environment as a whole. It is a system of knowledge that provides information to those who can read its signs and who are alert to its presence. It signifies awareness of the embodied nature of existence and the interpenetration of organisms in an environment that is not a container so much as a system of communication and network of relationships.

Conclusions

This is by no means an overview of /Xam, let alone San or African, concepts concerning the environment. I have gleaned only a few concepts from stories told 120 years ago. The more I think about the concepts I have chosen, the more I am aware of those I have left out. The sun, moon and stars, for example, all possess an extensive conceptual reach. The /Xam materials are weighted towards men's experience since Bleek had access to male prisoners and his efforts to obtain female informants were, with one exception, futile. Concepts drawn from gathering plants and rearing children, for example, are largely absent. Nevertheless, the examples I have given provide, I hope, an idea of the richness of / Xam, and by extension San, thought and practice in the context of a sustainable lifeway lived in a wild environment until the incursion of settler farmers into the area abruptly ended the viability of this way of life. The game was shot, the land fenced, the plants trampled by sheep, and the /Xam themselves murdered or turned into farm workers in the space of a few short decades. They live on, nevertheless, in the records collected in the Bleek and Lloyd collection, and in the presence and culture of their descendants. I believe that the /Xam experience and knowledge of the interconnections between the body and the environment, the living and the dead and the animal and the human, as articulated in their stories and the cultural information recorded by Bleek and Lloyd, have a greater resonance than ever in this time of environmental catastrophe and economic hubris.

Note

1 This reference and others like it refer to the unpublished notebooks of the Bleek and Lloyd collection. The letter L or B is used to indicate whether the notebook was compiled by Wilhelm Bleek or Lucy Lloyd. The Roman numeral refers to the informant. /Han#kass'o is identified by VIII.

Bibliography

Biesele, M.A. 1993. *Women Like Meat: The Folklore and Foraging Ideology of the Kalahari Ju/'hoan*. Johannesburg: Witwatersrand University Press.

Bleek, D. 1929. "Bushman Folklore." *Africa*, 2: 302–313.

Bleek, W. and L. Lloyd. 1911. *Specimens of Bushman Folklore*. London: George Allen.

Challis, S. 2012. "Creolisation on the Nineteenth-Century Frontiers of Southern Africa: A Case Study of the AmaTola1 'Bushmen' in the Maloti-Drakensberg." *Journal of Southern African Studies*, 38(2): 265–280.

De Prada-Samper, J (ed). 2016. *The Man Who Cursed the Wind and Other Stories from the Karoo*. Cape Town: African Sun Press.

Francis, M. 2007. "Explorations in Ethnicity and Social Change among Zulu-Speaking San Descendants of the Drakensberg Mountains, KwaZulu-Natal." Unpublished PhD thesis, University of KwaZulu-Natal.

Gordon, R.J. 1992. *The Bushman Myth: The Making of a Namibian Underclass*. Boulder, CO: Westview Press.

Guenther, M.G. 1999. *Tricksters and Trancers: Bushman Religion and Society*. Bloomington: Indiana University Press.

Hewitt, R. 1986. *Structure, Meaning and Ritual in the Narratives of the Southern San*. Hamburg: Helmut Buske Verlag.

Hodgson, D. 2009. "Becoming Indigenous in Africa." *African Studies Review*, 52(3): 1–32.

Hollmann, J.C. (ed.). 2004. *Customs and Beliefs of the /Xam Bushmen*. Johannesburg: Wits University Press.

James, A. 2001. *The First Bushman's Path: Stories, Songs and Testimonies of the /Xam of the Northern Cape*. Pietermaritzburg: University of Natal Press.

Krog, A. 2009. *Begging to Be Black*. Cape Town: Random House Struik.

Kuper, A. 2003. "The Return of the Native." *Current Anthropology* 44 (3): 389–402.

Laris, P., S. Caillault, S. Dadashi and A. Jo. 2015. "The Human Ecology and Geography of Burning in an Unstable Savanna Environment." *Journal of Ethnobiology* 35(1): 111–139.

Lévi-Strauss, C. 1964. *Totemism*. London: Merlin Press.

Lewis-Williams, J.D. 1981. *Believing and Seeing: Symbolic Meanings in Southern San Rock Paintings*. New York: Academic Press.

Lewis-Williams, J.D. 1997. "The Mantis, the Eland and the Meerkats : Conflict and Mediation in a Nineteenth-Century San Myth." *African Studies*, 56(2): 195–216.

Lewis-Williams, J.D. (ed.). 2000. *Stories That Float from Afar: Ancestral Folklore of the San of Southern Africa*. Cape Town: David Philip.

Lewis-Williams, J.D. 2015. *Myth and Meaning: San Bushman Folklore in Global Context*. Cape Town: UCT Press.

Lewis-Williams, J.D. and S. Challis. 2011. *Deciphering Ancient Minds: The Mystery of San Bushman Rock Art*. London: Thames and Hudson.

McGranaghan, M. 2014a. "'Different People' Coming Together: Representations of Alterity in /Xam Bushman (San) Narrative." *Critical Arts: South-North Cultural and Media Studies*, 28(4): 670–688.

McGranaghan, M. 2014b. "'He Who Is a Devourer of Things': Monstrosity and the Construction of Difference in /Xam Bushman Oral Literature." *Folklore*, 125: 1–21.

Mudimbe, V. 1988. *The Invention of Africa: Gnosis, Philosophy and the Order of Knowledge*. Bloomington: Indiana University Press.

Orpen, J.M. 1874. "A Glimpse into the Mythology of the Maluti Bushmen." *The Cape Monthly Magazine*, 9: 1–13.

Parkington, J. 1996. "'What Is an Eland?' N!ao and the Politics of Age and Sex in the Paintings of the Western Cape." In: Skotnes, P. (ed.) *Miscast: Negotiating the Presence of the Bushmen*. Cape Town: University of Cape Town Press, 281–289.

Prins, F. 2009. "Secret San of the Drakensberg and Their Rock Art Legacy." *Critical Arts*, 23(2): 190–208.

Schmidt, S. 2013. *South African /Xam Bushman Traditions and Their Relationships to Further Khoisan Folklore*. Cologne: Rüdiger Köppe Verlag.

Solomon, A. 2007. "Images, Words and Worlds: The /Xam Testimonies and the Rock Arts of the Southern San." In: Skotnes, P. (ed.) *Claim to the Country: The Archive of Wilhelm Bleek and Lucy Lloyd*. Johannesburg: Jacana, 149–159.

Spivak, G. 2006. "Subaltern Studies: Deconstructing Historiography." In *In Other Worlds: Essays in Cultural Politics*. London: Methuen.

Vinnicombe, P. 2009. *People of the Eland: Rock Paintings of the Drakensberg Bushmen as a Reflection of Their Life and Thought*. Johannesburg: Wits University Press.

Weintroub, J. 2016. *Dorothea Bleek: A Life of Scholarship*. Johannesburg: Wits University Press.

Wessels, M. 2010. *Bushman Letters*. Johannesburg: Wits University Press.

7

CAN THERE BE RELIGIONS WITHOUT BELIEF?

Religion in Latin America

Oscar Guardiola-Rivera

Introduction: only Amerindia can save the European soul

A few years ago, the Greek human rights philosopher Costas Douzinas and I were in Brazil for a series of talks and conferences. We had been listening to and learning from our Brazilian friends for a long while, at a time of great creativity and enthusiasm associated to the rise to power of Brazil's multiform Left, constituted by various social movements and self-organized around the Workers Party. All of them were true interlocutors whose different yet mixed outlook made all the difference – liberation theologians and philosophers, feminists and leftist thinkers, landless activists, playwrights, the successors of Third Cinema, young decolonials, human rights advocates, and especially Afro and Amerindian intellectuals. Between each and all of these singular voices we found brilliant thinking and committed militancy ready to engage, critique and create.

We were following in the footsteps of thinkers like Claude Lévi-Strauss, Michel Foucault and Boaventura de Sousa Santos, listening, learning and being transformed by this amazing country and its extraordinary people. Brazil's creolized thought, deeply rooted in the religious and cosmological memory of indigenous communities that are both constitutive and contemporary, rather than mere "remnants" destined to be sublated by progress and the violent onslaughts of colonization and capitalism, proved revelatory.

The crucial moment came when Costas began his final talk with a lapidarian phrase: "Europe is dead." The effect in the audience was very different from the one he expected. He could just as well have shouted, "Fuck you. God is dead." That would've been more acceptable. A member of the audience, a Brazilian gentleman of old age, stood up and before Costas could develop his argument firmly replied, "You cannot say that. Not here. We've spent five hundred years trying to think like Europe, be like Europe. It was the way of the future, you said; we were the past. We believed you. We believed in your belief. What now?"

The episode made the trip "interesting," as our friend Slavoj (Žižek) would say, neither raw nor well cooked, but rather, in the middle.[1] Slavoj's reference to Lévi-Strauss (neither raw nor cooked, but temperate) seems quite apposite in relation to the Brazilian episode. It not only relates to the problem of dual organizations, the question of binaries originating in religious belief as the supposed key to the totality of culture and politics, and its radicalization in the indigenous milieu of the Americas. It also invokes Aristotle's tri-partite geo-psychology, a crucial and half-forgotten element in the colonial origins of modern philosophy, global standardization and the body politics of the law also known as "philosophical geography," without which modern capitalism wouldn't have emerged.

It is the wager of this piece that the question of dual organizations and religious belief, radicalized in the indigenous thought of the Americas, merits renewed attention. If set in the framework of philosophical geography's now half-forgotten significance in the colonial imposition of concepts of the human upon indigenous peoples and their conceptual universe, the dual structure of religious belief appears as part of a geopolitical design. If so, we may have to reverse and invert our own conception of religion: to think of it less as the royal road towards understanding culture in its totality, and more in terms of the historically specific political, economic and legal elements that have come to constitute the reactionary basis of capitalism (as itself a cosmology) which still operate under camouflage in much of today's modern philosophy and politics (including some strands of criticism). To radicalize the binary structure of religious belief from the perspective of indigenous thought, in this case the thought of Amazonian and other Amerindian groups, means to engage in the practice of a permanent decolonization of modern/colonial thought as a project of global subsumption and standardization – a geopolitical design.

I refer in particular to the emphasis on the centre ground, the moderate and temperate middle. Also, to the (synthetic) middle or third way. This "third" way has been construed in accordance with the model of linear perspective, which proposes a sovereign viewpoint capable of systematizing timespace and everything within it, while at the same time positing itself outside of it – a focal "centre-point," or the site and state of exception. This "centre-point" is often conceived as the proper place of thought, the well-grounded foundation of "normal" psyche and self-affirming subjectivity vis-à-vis the "madness" of particularly disposed, nervous, supposedly excessive "primitives." Also, as the grounds for the resolution of alleged moral extremes and the taming of conflict.[2]

More generally, this visual metaphor, based on a stable line (the horizon line), hinges on the stability of an observer imagined on a ground of sorts, a boat, a coastline, as the faithful spectator of an event out there, a shipwreck, a crime. In this metaphor, the horizon line determines not only the limits of communication and understanding, but also acts as the very threshold of reason beyond which there's only muteness and silence, or the total liberation of instinct and appetite. Once imported into the realms of bio-psychology and geopolitics, or, later on, human and even post-human geographies, we rediscover this imaginary metaphor of the linear in the shape of the law of all or nothing: the ungrounded, deracinated and mobile "native" (alien, immigrant) in his or her inconstancy and madness, knows no limits.

First recovered between the two medieval and early modern scholastic schools (from Albertus Magnus and Thomas Aquinas to de Vitoria, Luis de Molina and Tomás de Mercado) by jurists and theologians, the moral authority of linear perspective and (the middle) place in philosophical geography operated as an organizing matrix of the lands and peoples of the Earth. From such a perspective they were viewed, as it were, from a distance, thereby establishing the peoples that occupied the temperate or middle zones as more advanced and the very standard against which the peoples of the torrid zones (the "tropics of empire") and the outer worlds or peripheries would be defined as backwards, and their psyche as feminine or child-like, excessive, hysterical or mad.

This way of seeing the world and its peoples (from above, at a distance, as the universe was once thought to be organized for the all-seeing eye of God), which functioned as an ontologically organizing and normative principle, obeyed a cosmological tradition that "that had long associated latitude with temperature, temperature with the nature of places, place with the physiology and psychology of peoples, and human nature with the 'place' that polities ought to occupy in a global world order."[3] Such a cosmological tradition, which included ancient and medieval intellectual sources, was re-elaborated, as you know, in the contexts of the Portuguese expansion into Atlantic Africa and later on to serve the needs of Spanish and other European powers' imperialism across the Atlantic.

Moreover, in the wake of wars of secession and independence against external colonialism, this way of seeing, imagining and calculating became the framework for the organization of post-war republics and economies around two central forms of internalization of the colonial: the subjectivity of persons (an inclusive-exclusionary, or "slicing and dicing" conception of humanity) and the objectivity of territories (a "natural order" of centres and peripheries, north and south, in or out of history). It is in recognition of this fact that those of us who participated in the seminal debates which launched the so-called "decolonial turn" in the 1990s but have remained suspicious of theories and theorizations that perceive themselves as non-ideological (post-modernism, post-colonialism, post-racial theories, etc.), nowadays tend to avoid speaking of "coloniality" and prefer to engage with questions of (subjective) *self-colonization* and *internal colonialism* as two sides of one and the same coin.

This "same coin," this framework – visual, ontological, normative – is the result of a mixture of old and new ideas that nourished and informed the intellectual mindset of explorers, conquerors, chroniclers, businessmen, lawmakers, contract theorists and missionaries at least from the 15th and 16th centuries onwards. Christopher Columbus and Amerigo Vespucci, André Thevet and Jean de Léry, Antonio Vieira and Hans Staden, Francisco de Vitoria and Luis de Mercado, Thomas Hobbes, Baruch Spinoza and the Scottish pioneers of political economy as well as their successors, their imagination and their political ambitions. This was the framework that set and staged their view of the south and drove them there in the first place. Thereafter, setting the stage for what would become a decisive moment of original acquisition and accumulation, the very blueprint for globalization, both constitutive and contemporary.

Put otherwise, this is the moment when the "persecuting society" that had developed within medieval Europe "nourished by militarism and Christian intolerance, which viewed 'the Other' mainly as an object of aggression,"[4] specifically in relation to "those who were regarded as deviants from standard Christian belief and practice," and the "demonization" of such groups,[5] on the one hand, and on the other, the more concrete need for a labour force of conquering Spaniards and other Europeans, became entangled. They fused to provide the frame and filter through which businessmen like the Florentine Vespucci, missionaries like Vieira or De Léry, jurist-theologians like Sepúlveda or Mercado, and the makers of the modern discipline of "Natural and Moral History" interpreted the cultures, religious cosmologies and sexual customs of the inhabitants they found in the Americas.

Cultural markers such as nudism, women's sexual freedom or their autonomy vis-à-vis community chieftains, polygamy and sodomy, which initially coexisted with more negative portrayals of devil worshiping and inconstant deviance contributed to the invention of the indigenous "Indian" as little more than beasts of burden. Together, such markers helped to configure the fiction that the conquest wasn't merely a financial enterprise, a rush for gold and silver, but rather a mission of conversion and civilization; and thus, that not everything about colonization was shameful.

Such claims, which in 1508 helped the Spanish Crown gain for itself papal benediction and absolute authority over the Christian Church in the Americas, are echoed nowadays by leading Oxford theology professors, conservative politicians and government officials, as well as peddlers of "prisoners of geography" discourse.[6] This time, in the fire this time, the argument is often advanced in the name of rights to freedom of expression and alleged evidence-led academic research rather than strict religious doctrine. Posing as dispassionate examinations of imperial projects, they attempt to balance out the violence committed to build empires with their supposed benefits, testing ethical/political critiques of coloniality against "facts" rooted in the self-serving justifications of imperial administrators with the aim of developing a "Christian ethic of empire," thereby rehabilitating yesterday's imperialism as a force for moral good in a way that can help justify the military and economic interventions of not only yesterday, but also of today.

Notice that in both cases, yesterday and today, the "centre" holds: the equation consisting of belief (grounded, centred, temperate) + Christendom = moral good. Belief is understood here as faith in the thaumaturgy of the immaterial, be it soul or mind, conceived as a potency that may transform the world from within. Such faith is based on the supposed intrinsic power of knowledge and values as pure public goods that arise and may be accumulated in the centre, and then transplanted elsewhere to exert their natural productive capacities. This sort of analysis not only conveniently forgets that the productive capacities of values and knowledge aren't naturally determined, but rather, that social norms and legal rules performatively instituted and sustained limit or not the fecundity and extension of knowledge. But also, that in the conditions of modern coloniality and capitalism, this happens in terms of the organization, control and orientation of global and collective work, that is, in terms of domination (Christian domination).

As in the 16th century, these claims make invisible to the eyes of the world, and possibly to the colonizers themselves as well as their successors, any sanction or proper remedy against the beneficiaries of the atrocities that they may have committed or continue to commit against Amerindians, Africans and other "southern" peoples. This way of (not) seeing works "as a license to kill, independent of what the victims could do," as Silvia Federici says. She observes how in spite of early sermons about the kind Taínos or the good savage, "an ideological machine was set in motion, complementing the military one, that portrayed the colonized as 'filthy' and demonic beings practicing all kinds of abominations, while the same crimes that had been attributed to lack of religious education" were subsequently treated as signs of cognitive and ethical lack or physio-psychological deficiency. Amerindians were associated to the alleged nature of their environs – excessive and exuberant like the jungle, in contrast with the stone-like certainty of belief of the nations placed in the temperate zones of the Earth. Set in such a manner, this binary immediately evokes notions of certainty of belief versus threatening inconstancy and risk. Such images of the "torrid zones" would easily evoke in the minds of European observers familiar pictures of hellfire and damnation, and therefore appear as evidentiary signs "that the Indians were under the dominion of the devil and they could be justifiably deprived of their lands and their lives."[7]

Hence the importance of asking the question I hope to examine in the rest of this paper in relation to the reversal of Western religious binary structures in Amerindian indigenous thought and cosmology: can there be religions without belief? In what follows, I'll proceed by means of a brief genealogy of the religious creolization that took place in the Americas in the wake of the encounter between Christians and Amerindian "Cannibals" as well as Africans.[8] I'll emphasize questions of method in relation to the transcendental categories of religion and the undoing of belief, as well as its consequences in terms of a philosophy of action, law and politics. In doing so, I'll both supplement and criticize our modern onto-theological reliance on the synthesis of binaries in the unifying third, "simple" good/evil binaries, and the normative authority of the middle place as the immanent point from where a transcendental (immaterial) power can transform the world and inter-human relations, serve as "a force for good" due to its alleged intrinsic capacity, as well as a "sheet" that can be naturally balanced. My (genealogical) standpoint will be the perspective opened up by the dialogue that began to take place in the late 1960s between feminist theory, anthropology and Amerindians (the "spirit of 1968" in the Americas).

A brief genealogy of religious creolization

For the purposes of this brief genealogy, we can recognize in Christianity a body of knowledge about the human world, one that is a principal source of inspiration of the human sciences, but crucially, one that does not relate to other bodies of knowledge and values from a position of unique hegemony. In this framework of creolization, or "the association of associations of various classes and social" as well as political identities, militant agents may well include in their actions elements that are recognizable as "intellectually Christian" working together with indigenous (Afro and Amerindian), peasant and workerist ones.[9] This actional framework is a result of the opening of scholastic thought to the historical conditions of its encounter with the explosive conditions of Amerindian cosmological and religious thought in the Americas. In time, a new religious and philosophical dimensionality would emerge out of this milieu informed by the Afro-Amerindian cosmologies that populate the world of the poor in our part of the world. It may be recognized not only as a heterodoxy but actually as a worthy challenge against the "creeping neo-Christendom" of Christian theology.[10] Known in the wake of the 1968 Medellín Conference as "Liberation Theology," it's also superior to most theological orthodoxies in that it's the only one that inspires radical action through its prophetic voice and incarnational vision: "real incarnation in this world . . . leading to the cross" and putting death to the service of life, in Jon Sobrino's terms.[11]

In this respect, and in the historical context of the struggle against colonization, Amerindian religious cosmology could be read not only as a "Death of God" a-theology, but also as a cosmopolitical perspective in which "human beings proper are a species within a multiplicity of other species of human beings who form their own societies." In other words, a generalized personhood. An implication of speaking of the "own societies" of meta-person others among the Araweteór the Achuar is that they do not conceive of "the supernatural as a level of reality separate from nature," insofar as the human condition is common to "all nature's beings. . . . Human beings, and most plants, animals, and meteors are persons" with a soul and individual life. Divinity originates as a kind of animism of higher taxonomic order, with metapersons as participants in and controlling forces or ordering principles of life. Rather than economic or legal subject/object relations of production, in such a system, women and men are recipients of the plants they cultivate, have a personal relationship with them, talking to and helping them mature. Similarly, in this cosmo-political economy, laws and commands are the result of humans absorbing divine powers, and those humans will soon find themselves confined to the (utopian) space of the sacred and fought against if they try to become a ruling class.[12]

Such radical anti-institutionalist politics and fighting universalism informed the high scholasticism of people like Bartolomé de las Casas in the 16th century and 20th century liberation theology and philosophy. Instead of the previous binary distinction between God's realm/this world, the problem of the bifurcation of being is reset in a way that refers back to a distinction between life not itself alive (that is, the form and constancy of the soul according Aristotle's most post-Platonist work, *De anima*) and living beings. Arguably, what comes in the wake of the encounter between European Christians and indigenous Cannibals after the 15th century is a new and different answer to the problem of multiplication and the order of beings: how does one bifurcate into two? How does one come to be (out of) two? How are we to make sense of this gap between creator or producer and the recipient produced? As can be seen, this matter pertains to life in the most concrete sense: on the one hand, to sex as both transmitter of life and disjunctive division (*sectus*); on the other, to the pathologies of freedom that emerge as soon as we realize we are no match for our creations, especially in the technological world – the technosphere that comes in the wake of teletechnologies of linear vision applied to the navigation, conquest, colonization and capitalist extractification of the globe – from maps to the stack.

As is known, the latter term in the Aristotelian distinction described previously – living beings – is further differentiated into the three levels of the soul. These are three different levels in which living beings, especially but not only humans, can be receptive to the soul's forming power (i.e., light, reason) to the extent allowed by the physical or bodily properties communicated by place in accordance to the neo-Platonist reading of the distinction between cold Northerners, hot Africans and South-Asians, and temperate Greeks as well as Europeans in Book VIII of Aristotle's *Politeia*. This tri-partite model, itself indebted to the description of the human soul as constituted by three hierarchically related parts in Plato's *Republic* – appetite or instinctive perception (*epithumia*), spirit or emotion (*thumos*), and calculation or reason (*logismos*) – would provide not only a psychological but also the geo-political template for the encounter between Christians and Amerindian Caribs or Cannibals during and after the 15th century.

Aristotle's geo-political model accords to middling Greeks ("Greeks carrying gifts," as per the story of the Trojan horse in *The Iliad*) the ability to carry out reason's works by summoning willpower against the predatory claims of unrestrained instinct. Barbarians, in contrast, either fail to reason properly or cannot summon the requisite strong willpower; they give in to laziness and inconstancy. Despite its apparent simplicity, in the hands of Christian jurists and theologians such as the Scottish Dominican John Mair and the Crown chaplain Bernardo Mesa, this model has proved an enduring rationale for Western colonialism and totalizing racialist anthropologies, which posit religion and belief as the royal road to the essence and ordering of being and culture. It assumes constant belief (in God or the tribe and its king, with reason assisted by spiritedness) as the impulse to contemplate and constitute the whole, to restore oneness, and thus as the being and the perseverance in being of the group as a reflexive and identitarian totality.

From the outset, Amerindians and their allies questioned the centrality and totalizing character of belief, casting suspicion on the idea of society as a calculated totality founded on the obsessive policing of an essential hard core and its limits, which uses the exterior as an inverted mirror for self-identity. Put simply, Amerindians were suspicious of their Greek-Christian fellows bearing the gift of totalizing belief or "true faith," much like the Brazilian gentleman of Costas's story.

It's not that they refused the gift. The fact is they accepted it. However, being experts in the alter-law of symbolic exchanges and reciprocity,[13] Amerindians recognized it for what it really was. Namely, an empty gesture. For they were being asked to embrace freely what was already being forcibly imposed on them. In a move that enthused their conquerors, Amerindians accepted the Europeans' true faith. But they would welcome God's word and kingly order one day just to ignore it the next. Much like the Brazilian gentleman of Costas's story.[14]

Put in relation with the question of method in the early modern constellation, the Amerindians' move can be understood thus: to deal with belief, true faith, law and life or being as in some sense catapulted to the level of a totalizing, transcendental ontological category or structure. Going further, to arrive at the bifurcation or multiplication at the heart of this totalizing structure – which undoes the structuring order from within. Furthermore, the invitation is to move from the transcendental divide of the one intrinsic power towards multiple speculations and disobedience to any external force imposing itself upon the collective work and reasoning of all.[15]

From bifurcation to multiplication: religion, liberation and permanent decolonization

The move from bifurcation to multiplication in matters of belief via a method of quasi-scholastic transcendental inquiry is the very mark of liberation theology and philosophy, as well as all philosophical anthropologies of permanent decolonization. This is the case in the tradition that goes from the "defender of the Indians" Bartolomé de Las Casas's arguments in *De unico vocationis modo* (1530) to Jon Sobrino's incarnational vision (1992), Enrique Dussel's *anadialectical* transmodern method (2003) and the *Cannibal Metaphysics* of Eduardo Viveiros de Castro, all of them informed by Amerindian religious cosmologies.[16] The point of liberation philosophy and the anthropological theories of permanent decolonization is that this fundamental multiplication must itself be thought within the framework of the totalizing belief or structure – for instance, the providential messianism of Europeans, which narrates historical events in the context of a universal confrontation between true faith and infidelity – which it simultaneously undoes.

This is to say that in accepting the Greek-Christian (religious) gift but choosing not to believe, Amerindians and their successors effectively challenged the religious ideological matrix of their Christian conquerors. In such a matrix, the image of the border separating the centre from the other parts of the world (and the soul) is read as a line of separation between "us" and "the enemy," meaning that war (of extermination) plays a strategic role in the definition of the forces between the groups and peoples inhabiting the world by rendering them defenceless and advancing the "desertification" necessary to the Christian capitalist colonization of the exterior.[17] In Žižekian terms, Amerindians sought to disintegrate the semblance of freedom inherent to the symbolic and geo-psychological order that emerged from the Greek-Christian gift offered by European religious missionaries and jurists.

Amerindians acknowledged the fact that the choice being offered by the Europeans was, like every choice we confront in language, a choice of choice itself; that it involved the question of belief in belief, or as Jean Pouillon would say, that there is always doubt at the heart of trust and conviction so that to believe is to confront that contradiction at the heart of one's faith. Moreover, Amerindians put forward the thesis that this was the result of distinguishing between two worlds – "this world" or nature, on the one hand; and on the other, the "not this world" of mind, culture and supernatural religion – and sought to test such thesis.[18]

In both accepting and refusing the Greek-Christian gift, Amerindians made a choice that effectively changed the coordinates and object of their choosing. To perform an act that changes its very object, "the deliberate directing of . . . energy in action that . . . fashions new relationships and new reality," distinct from so-called free choice, is for liberation theologians what can be properly called freedom.[19]

Now we can understand how is it that in encountering Christian Europe, Latin Americans can be eagerly pro-European and fiercely anti-Eurocentric at the same time, and further, eagerly anthropomorphic and furiously anti-anthropocentric.

Let's now take a further step in our brief genealogy, moving to the story of what happened when Christians encountered the indigenous Carib, which they renamed Cannibals, and the latter engaged in an experiment concerning the possibility of religion without belief.

The encounter between Christians and Cannibals

On the fourth Sunday of Advent (21 December) 1511, after having enumerated what he called the *crueldades exquisitas* committed against enslaved Amerindians, Dominican friar Antonio Montesinos informed the stunned members of his audience in the island of Hispaniola that the way things were going they could "no more expect salvation than the Moors . . . who lack the Faith of Christ and do not wish to seek it." The outraged audience included not only one of the future founders of liberation theology, Bartolomé de Las Casas, but also His Majesty's governor Diego Colón (Christopher Columbus's son).

Promptly afterwards, the settlers appealed to King Ferdinand the Catholic, by then the widower of Queen Isabella I of Castile whose main concern was to maximize extraction of gold and silver in the colonies, and by implication, the need for ever-greater quantities of forced labour. In their legal appeal, the settlers construed Montesinos's diatribe as a challenge to the Crown's sovereignty, and a declaration of war against the King's title over the New World. This obliged King Ferdinand to call a gathering, or *junta,* of jurists and theologians to meet with his councillors at Burgos in 1512 in order to examine the legitimacy of his title in the Americas.[20]

The Burgos *junta* concluded the Indians should be treated as free subjects of the King and could be ordered to work as long as their labour did not interfere with their conversion, and as long as it was tolerable labour. Unhappy with the findings, King Ferdinand and his settler-investors asked for a second opinion from the Crown's chaplains. One of them, Bernardo Mesa, wrote in his *parecer* or legal opinion that the Indians weren't slaves *de iure* (since they hadn't been won in a Holy War against infidels and could not count as infidels themselves) neither were they slaves bought in the market nor born of other slaves. The only reason he found was "their lack of understanding . . . and the absence of will to persevere in the practice of the Faith . . . for that is the condition of natural slavery, as the philosopher says."[21]

The philosopher in question was Aristotle. Mesa added, referencing Aristotle once more, "or perhaps they are, as the philosopher says, slaves on account of the nature of the land, because there are certain lands that the aspect of the heavens renders subject to others." Although this was not the first time the doctrine of natural slavery was cited by European imperialists, it was certainly the first time that an official debate included the question of the so-called "inconstancy" of the soul of the Indians (the absence of will or spirit to persevere in the practice of the faith, which they had "reasonably" accepted) as a problem that could lead to questioning the legitimacy of the King's title in the New World, and thus to war. Crucially, the problematic nature of the Indian soul was linked in these debates to the alleged moral authority of place and geography. This was the first time that apologists of empire effectively argued the insignificance of the global south.[22]

In his 1550 *Apologia pro libro de iustis belli*, a summary of his better known *Democrates secundus* (1544), the Crown jurist and translator of Aristotle, Juan Ginés de Sepúlveda, cited the tri-partite geography of races or nations which invoked the Aristotelian diagram of the three souls to claim that Indians did not meet the natural preconditions for preserving political autonomy from Christian Europeans. A quote from Thomas Aquinas's own commentary on Aristotle's *Politeia* permitted Ginés to render the issue in terms of the alleged inconstancy of the Amerindian soul. Barbarians, he said, are "those deficient in reason, whether because of the region of the heavens, which makes them weak, for the most part; or because of some evil custom, which makes them almost like beasts." As Las Casas put it after his confrontation with Ginés at the Valladolid *junta* of 1550, this learned articulation of geo-psychology and politics constitutes the most detailed and comprehensive blueprint for imperialism.[23]

Recall that in what used to be called Hispaniola, today's Haiti, the native population, which numbered about 100,000 in 1492, had dropped to 200 a century later. Many of them had died "of horror and disgust at European civilisation even more than of smallpox and physical ill treatment," to take ours Lévi-Strauss's words. "Commission after commission was sent out to determine the nature of the inhabitants" in the wake of the *juntas*. If they had souls "and were really men, could they be descendants of the ten lost tribes of Israel? Were they Mongols who had ridden there on elephants?" If so, where did the elephants go since none were found? Or were they "Scotsmen brought over several centuries before by Prince Modoc? Had they always been pagans or were they lapsed Catholics baptized once upon a time by Saint Thomas?"[24]

Some suspected the Amerindians indeed had a soul, but an inconstant one, given that, as the Jesuit Antônio Vieira put it another century later,

> they receive everything that is taught to them with great docility and ease, without arguing, without objecting . . . without resisting. But they are statues of myrtle that, if the gardener lifts his hands and his scissors, will soon lose their new form, and return to the old natural brutishness.

Vieira concluded that this people with no faith, no law and no King "did not seem to offer a psychological and institutional ground in which the Gospel might take root."[25]

Of all these commissions, the most deservedly famous was that of the monks of the Order of Saint Jerome. In the course of what was "tantamount to a psycho-sociological inquiry, conceived according to the most modern standards," as Lévi-Strauss says, "the colonists were required to answer a series of questions, the purpose of which was to find out if, in their opinion, the Indians were or were not capable of living on their own." All the replies were negative, of course, on the basis that the natives carry

> perversity to the point of giving away their possessions. . . . Evidence given a few years later adds the finishing touch to the indictment: "They eat human flesh, and have no form of justice. . . . They do not have beards, and if by chance hair grows on their faces, they lose no time in plucking it out."[26]

These were "fact-finding" commissions in the sense in which the term has come to be used nowadays by political philosophers, international lawyers, experts on geopolitics, IMF economists and human rights do-gooders. On the basis of Aristotelian geo-psychology and Roman law, following the model of inquiry inherited from the *concilium* and the *inquisitio* in places like Burgos, Valladolid or Coimbra, and thereafter in Paris, London, Oxford and elsewhere, these experts attempted to solve the legal and theological problems posited by the existence of so-called naturals and their cosmologies in the Americas.

Notice that these were unfamiliar problems; for the enemy here was not a different orthodoxy or dogma, which they could recognize as a lower-ranking belief system in the hierarchy of beliefs, but rather, indifference to dogma and a refusal to choose to believe.

My contention is that the severity and persistence of this "problem" has had to do with the inability of standard normative historicism, Christian and otherwise derived from the model of Christendom, to come to terms with the challenge, a war-like challenge, implicit in the injunction to live with others and to learn how to live with others without a Big Other. This historicism has become more pronounced in our time, even after the various "linguistic" and deconstructive or rhetorical turns, as standardization, persuasion (or confession and conversion) and the primacy of conventions take root in the humanities, law, psychology and the social sciences, as the lines quoted here by Lévi-Strauss suggest.[27]

Standardizing historicism and conventionalism are accompanied by a much too poor understanding of performativity and rhetoric, and is firmly entrenched in the model of the self as economic man that developed in parallel with the emergence of global dominion, flows of trade, legalist contractualism, communication and financialization between the late middle ages and early modernity, supported even further by the late modern digital (counter) "revolution" of recent years. This is what the Amerindian refusal is all about. For (Christian) historicism makes true belief (not just *pistis*, or trust, but *kerygma*, or proclamation, precisely speaking) not only one belief among many, but also and at the same time, allegedly, the one, unique or "unical" belief, hierarchically superior to all others. Similarly, persuasion and confession come to be treated as the expression of the sources of legitimacy of an organized compact or public in its "shared" values or conventional rules.

The (Amerindian) problem with Christian normative historicism

In brief, Christian historicism, which is a direct result of the admixture of imperial legalism, Greek-Christian geopolitical psychology, and the sort of psycho-sociological inquiry carried out in the Americas according to the "most modern standards," proclaims the doctrinaire teaching about Jesus an aprioristically dictated, persuasive, conventional, posited source of legal legitimacy and a guaranteed (pre-destined) blueprint for action. In doing so, it makes the person-event that lies at the heart of the specific notion of Christian belief (i.e., resurrection/conversion/confession), or, to be more precise, the link between the particular incarnating person and the incarnated event or absolute experience, a necessary one. Notice, however, that if the link between the incarnated event or the absolute and the incarnating person or content becomes necessary, then the former would lose its dimension of transcendence, universality or beyond – its meta-personality, to make ours the language of anthropological explanations of original political society. Having lost its dimension of transcendence or meta-personality, the universal is subsumed by or absorbed in the incarnating particular, thereby becoming merely a token of some invariant type or necessity. In that case, "one would be able to name God directly," transforming it into an invariant type, a need, binding the god-figure to its function in response to our needs and ourselves to his representative authorities. Isn't that, precisely, what the term "religion" (*re-ligare*) has come to stand for? Precisely, a civic cult. In other words, the one authority or the few may claim to have discursive and henceforth also ethical, legal and political "mastery of His essence."[28]

This is what occurs when, for example, formulary statements of damnation, condemnation, curse or anathema not only function to define the creed or to distinguish infidels, but moreover, to assume an organized state-like authority to deal with others and establish obedience to it, thereby becoming law-like in the sense of that term which is more familiar to us as subject-citizens of a world of rule-of-law states. In other words, the few come to rule over the many as the universal exception to the rule.

To be more precise, they, the one and the few, come to decide on what is to be taken as the exception and thus elevate themselves to the place of a universal predicate (constancy of belief, the will of all). This is done through a procedure of aggregation of approaches linked to otherwise divergent camps, groups or individuals, and their needs, interests or investments that fix their attention on a specific text or object acting as the singular node for organizing identity. Consequently, every relation with others becomes an object relation. Whereas the space of shared meanings in a cosmopolitical relation such as that produced in the ritual practice of the (Amerindian) original political society is the spacetime interval for active negotiation, the space of the object eliciting overlapping attention (belief, faith, consent) from divergent groups or individuals evacuates the conflictual negotiation of meaning and orientation between subjects.

In such an emptied space, labour comes to refer to the subject's production of the object as opposed to the object's mediation of the (receptive) labour between subjects. This is to say that what is at stake in such relation is no longer the intrinsic personhood, importance or uniqueness of the other, but rather, the ways that the exceptional subject figures the other (all others) as object(s) for its own purposes.

The result is, on the one hand, a practice of identity production, inseparable from imaginary cultural or conventional elaborations which isn't just symbolic, but also makes symbolization into a site of self-affirmation and grounding. In other words, this phenomenon may be better understood as a politics of narcissistic projection and recognition in which practices of confession, proclamation and other rhetorical forms come to be conceived as a mode of inter-passive aggregation of overlapping approaches rather than reciprocal inter-personal or inter-subjective negotiation. An idea of "public utility" begins to take shape here, based on the older emblem of pastoral power – the shepherd and his sheep – but, crucially, one that determines such a relationship in a completely new way: one that reduces the plurality of the *sensibile* to mere ephemera and then submits such plurality to the unity of needs and the useful (utilitas) first, and, thereafter, to its exchange and expository value. That is, as useful objects for a subject. The usefulness or value of such objects, which may and have historically included human and other natural bodies, comes to an end in the very act of their consumption by a subject. Thus, we may speak of value as connected to their disposal or disappearance.

On the other hand, the public or people that emerges here as a social bond and boundary is only possible under conditions of specificity and particularity. Whether such conditions take the shape of a set of rules about how to relate to strangers (perhaps modelled on the Roman *ius gentium*), or a set of procedural rules and standards governing proper modes of mutual recognition as part of the public (on the basis of the presence of and constancy of will, taken as a psycho-geographical constant), the rendering of these conditions into the terms of the late *Respublica Christiana* (first, and later into so-called *ius publicum europeum*) entail conditions of exclusion, disposal or disappearance. For some modes of textual investment and procedural recognition to be meaningfully public, they must logically exclude other social links.

Thus, for instance, the *Respublica Christiana* that emerged on the basis of Christian political theology in the middle ages and early modernity established rules for the annexation of new territories with implications for those losing their grounds. In accordance to common law (*ius commune*, grounded on the common usage of Latin and legal as well as religious texts rendered in that language), Christian soil was annexed as equally important (*aeque principaliter*), whereas conquered heathen soil would be the object of *incorporación accesoria* or accession, as per the doctrine set by Guillelmus de Cugno and then taken up by Bartolus, Baldus and the Spanish school.

The kingdoms and ducados of Toledo, Navarra, Córdoba, Granada, Flanders and Naples, among others, were annexed *aeque principaliter*. In contrast, the Western Indies or *indias occidentales* of the Americas were the object of an accession. The former were recognized as bounded selves with cultural conventions, *fueros* and laws of their own, whereas the latter were absorbed within the self of the conqueror and the conquering state, its laws and by force. Both entail forms of inclusion and exclusion, but only the latter is at the same time both integral, eliciting affective identification because the conditions of inclusion would help in managing the problems of groundlessness in the wake of conquest, and of being a separate self by providing a set of narratives about who the subject is, as well as practices of self-referential relations with the other from an exceptional standpoint, establishing a particular kind of asymmetry.[29]

Here, investments at the level of the specific economy (bounded selves and self-referential relations grounded on textual forms) operate as a suture for the failed unicity of the more general economy (the ordering of beings proclaiming the unical authority of a central place, its faith, king and law). The conditions entailed by such a form of integral exclusion in a public generate a mode of enjoyment "that stems from embracing a violent dis-symmetry against others as a proxy for the problems of the general economy." Moreover, this logic is made all the more efficient if (its violence) is repressed and erased from the historical record.[30]

Thus, when revisionist historians like Elvira Roca Barea argue that the Américas weren't colonies because the specific character of their legal annexation meant there was no political subordination but total integration (legal, institutional, religious) with equal rights, or that the word "colony" wasn't frequently used until the accession of the Bourbons to the throne of the Spanish/Holy Roman Empire in the 18th century, and that Spaniards saw themselves as continuing the enterprise of the Reconquista rather than engaging in a business operation (in contrast with the Dutch, the French or the English), they are effectively erasing the facts of exclusion, as these were actually and differentially experienced by the various aborigines peoples.[31]

Not only that, they're also bolstering the fantasy of the social as a space of symmetrical and harmonious relations without violence. If this form of cultural annihilation isn't tantamount to genocide, at least not yet, it is at the very least continuous or derivative in relation to it. Its value depends on the repression of the underlier (violence), the over-appreciation of hypothetically shared values or the benefits of a supposed uniquely "civilizing" influence, and extracting pleasure (from anathematized others) in the (communicative) elevation of the contract-like communion of signs.

The repression of the violent impossibility of the (rhetorical, social, political) relation and of recognition of others through self-affirmation and mere aggregations is what makes up various forms of external and internal colonialism to this day. It's also what connects these more or less modern forms of internalized colonialism with older designations of others beyond the horizon of our perspective as mute, abject or as objects incapable of making objections of their own and engage in combat or stasis – the kind of constant civil war that anthropologists, analysts and their interlocutors see in practice as the exception to the exception and the universal predicate (that there's no ultimate ratio and communion is missing).[32]

The decisive event in this respect is the Council of Nicaea, convened by Emperor Constantine the Great in 325 AD with the aim of unifying the various tenets of Christology once, the empire having reached its limits, its rulers (first Licinius and then Constantine) embraced the movements of popular contestation that became remembered with the name of this great religious and philosophical tradition in a sort of last-ditch attempt to restore and maintain their rule. In Nicaea, the assembled bishops were asked to accept a statement of teaching set out in the form of a declaratory creed.

The main point of such a statement was to exclude the teachings of the Libyan ascetic Arius, his mentor Lucian of Antioch *ad invidiam*, and his followers who disagreed with the doctrine of coequal Trinitarianism and imagined a time before the Son of God. After Lucian and Origen, Arius argued that the created Logos was not only subordinate to the Father in the order of beings, but had also had a beginning. This means the Son/Logos was created and a recipient of the rest of creation made by the Father, of a different substance in relation to the infinity and eternal character of the Father as a meta-person. If put in terms of historical development, this would mean that only the Son or Logos is historically determined and his coming a historical event rather than the result of necessity or a unilinear outcome of Christ's being co-eternal and consubstantial with the Father. The declaratory creed upholding the Christological teaching according to which the Son is co-eternal with the Father would mean, conversely, that the Son's coming isn't a rupture but a mere development or continuity.

Decisively, 218 out of 220 bishops attending the Council signed their acceptance. An indication of the change in function of this, the first of the conciliar creeds, is the fact that the document drawn establishes, as if written in stone (as "the will of all," to use other terms somewhat anachronistically) not only what is the received belief, but also what is not. The latter is anathematized. *Anathema*, a curse, was pronounced on those who held Arius's propositions. Hereafter, belief comes to define not only the Christian in respect to the non-Christian, but also to distinguish the True Christian from the false (the heretic) in a manner that foregrounds later, modern friend/enemy distinctions that would be distinctive of state relations, internal or external. As in external or internal forms of colonialism and the kind of historicism which underlies it, or, more recently, in external or internal counter-insurgency decrees and operations.

My point is that not only the friend/enemy distinction is itself historical and contingent upon the appearance of state-like organizations. But also that the enemy is represented as threatening to disrupt a line of historical development assumed as the most basic aspect of human existence. Which means that the friend/enemy is not ontologically primary and should not be taken as constitutive of the political as such. At best, it's constitutive of the "modern" political if we understand the latter to mean only a politics of nations or publics emerging down the line from the great empires or religions, assumed to be determined by their more or less shared or homogenous appearance and values, constant loyalty before decreed rules and loving (fraternal, friendly) integration with the unique god-like ruler.

In contrast, I argue that what is primary is the antagonism between the ruler who seems himself as god-like and legitimized in our love for him, a purely self-referential relation, and the people who emerge as such only in the context of *stasis*, lodging an (heretical) objection against the ruler's appeal to exceptionality, and engaging in protest or civil war. A corollary of my argument is that *stasis* – heresy, objection, or civil war – is primary vis-à-vis international war.

The Declaration of Nicaea invented the heretic, a figure of the stranger that didn't exist before insofar as its condition of possibility is integral exclusion: his exclusion is necessary for the perseverance of Christian society in its own being. His exclusion restores the past line. War against the heretic belongs to the order of restoration and *aletheia*; that is, return to origins and social reproduction. This is also the order of the instituted and reconstituted, politically and legally speaking, or a decisive effort seeking the restoration of (our) being against the contaminating presence of (their) external becomings. From here onwards, wars of reconquest, crusades and, later on, wars of colonization, evangelization, and the civilizational mission could be justified by appealing to such a figure.

Moreover, the Declaration's "function [stating what must not be considered as the received belief] assumes an organised authority: bishops in council and not just in their sees."[33] In this sense, the heretic is not just a stranger, but also a stranger at war with the organized authority of a state-like institution and its posited rules. Not just an enemy, but *the* enemy of all. He is the image of external or internal enemies disloyal to "our" values not only because he might resist them in the name of some relative values that would be grounds for doubt and objection, or because he takes up much effort before he gives himself over. This would mean that, conversely, "once they have received the faith, they stay firm and constant in it, like statues of marble."[34]

Instead, the heretic isn't merely able to choose, but in a sort of double-twist on the etymological Greek root of the term, he can choose and yet he does not. Like Arius, he's inside and outside, both at the same time. He's double – diabolical in the most literal sense of the term. But in inverted form, he's the conceptual persona of the subjectivity engaged in a war of liberation, not a war against infidels or outsiders, but a war against "our" own declared authorities.

The circumstance of the Declaration of Nicaea was associated with the patronage of Emperor Constantine, on the one hand, and on the other with the acclamation of Constantine as Christian Emperor. It gave place, as its obverse, to the malediction of Arianism. As is known, the Arian controversy was never solved on its merits, but in terms of community loyalties and identities (chiefly, the major cleavage between East and West and the relationship of the Church and the Roman Empire). In this way, the demonstrative method of Greek philosophy (especially pre-Socratic philosophy) was pretty much abandoned and replaced by the legalistic and psycho-sociological method of declarations, acclamations, inquisitions, and anathematic or other such legalistic judgments. This decisive shift was to have momentous consequences for the development of medieval and early modern philosophy and politics, as well as for the anthropocentrism (and subsequent ethno-centrism) of the human sciences.[35]

Not only does such use of belief actually turn the believer into its opposite, the unbeliever, somewhat paradoxically. It is also the case that in fending off the threat of (moral, ethical) relativism, which it itself creates, the community of believers, or more precisely its authorities, come to occupy the dual position of being hierarchically superior to all others (like a super-species, or the only species) and yet one among such others (just another species). As the judge of/against relativism, the "vertical" ethics and moral vision of the community of believers is placed *sub species aeternitatis*, rendering it more or less incapable of helping in the construction of ethico-political, economic, historical and geological life, which as we all know, takes place in a terrain that is less than eternal.

It is precisely against this sense of "legalized resurrection" as creed, as both historicism and a priori necessitarianism, that liberation philosophers like Enrique Dussel, Jon Sobrino or Ignacio Ellacuría, inspired by the thought and struggles of Amerindians, have directed their challenge. They do so in precise terms. For they speak of and denounce "creeping neo-Christendom" or the becoming empire of Christianity, while emphasizing the dimensionality (i.e., time-changing engagement, utopian vision and prophetism, or *kairós*, in Enrique Dussel's parlance) of the incarnational vision of the theology of the cross. This is a "radical drawing near for love and in love, without escaping history or manipulating it from outside, wherever it leads" even unto death, death put in the service of life, to use Sobrino's terms.[36] It is indeed the closest thing to a call to arms against the very model of historicism and modern empire.

On the disobedient transformation of Christian normative historicism

This isn't just a matter of distinguishing between church authority and individual reason. Rather, it has to do with the distinction between belief as declaration and belief as commitment, the extra-dimension added by Luther when the latter renovated the explanation of belief as personal commitment, already implicit in Paul's *Letters*. Importantly, this distinction has organizational implications but relates mostly to the psychic drives. The distinction here is between the law, not only as outward performance but also as internalized obedience, on the one hand, and on the other, grace as action in solidarity or "effective mercy" in the language of Jon Sobrino.[37]

Here, one may draw a comparison with the Heidelberg Theses: "The Law says 'Do this' and it is never done. Grace says 'Believe this' and everything is done." The comparison stands. In fact, this wouldn't be the first time it has been pointed out that liberation theologies and philosophies made the Americas "the land of the second Reformation."[38] But it would be more appropriate to speak of this reformation as the first to occur on American soil. Not only because, as Joerg Rieger says citing Dietrich Bonhoeffer, "North American Protestantism has always been a 'Protestantism without Reformation,'"[39] but also because Amerindians, performing as inconstant heretics engaged in a war of liberation, were the first to drive a wedge between the law and engaged action in that continent. In doing so, they provided the basic motif for liberation theologies and philosophies as well as anthropological theories of permanent decolonization. Notice too that engaged action is precisely what's at stake in the idea of the "protection privilege of the poor," as Frederick Herzog would say, in a way that has nothing to do with post-modern appeals to "religious" or "cultural otherness," or with a Jesus subservient to some universal framework of religion. Rather, it refers to God's walk among the suffering, a moral image that invites us to commit to others in distress and struggle with them, yet not for them alone, concretely and subjectively, without reservation.[40]

In this respect, we can go even further and say that Amerindians are the paradigm of the subject who possesses belief by actually being possessed by it, as in the case of the inconstant Indians of the 16th century or the jaguar shamans of the Amazon and elsewhere; more clearly so than in the case of Luther himself. Let the clash between Christians and Amerindians be understood in this precise sense.

For in refusing to confess and believe, Amerindians were in fact refusing to obey the law (of the empire) in order to engage (to act in solidarity and reciprocity, to walk among the suffering, to think otherwise) so that, literally, everything could be done. This is what the Amerindian maxim to "multiply the multiple" and the Perspectivism of their religious cosmologies and philosophy stand for.

Later on, we may return to this refusal and its impact on the paradox reduplicated in our time by the (digitalized) human sciences: that it is the unbeliever who believes that the believer believes.[41] For now, it will suffice to notice that the encounter between Christians and Amerindians was not a clash between two self-serving dogmas or ideologies, one more universal than the other, but a clash of cosmologies, a clash of universalities, both of them equally totalizing, but only one of which would explore the possibility of perceiving otherwise a seemingly mutually exclusive situation.

This perceiving and thinking otherwise does bear comparison with the psychically driven meaning of belief in Luther, for whom the "perceived presence of the Word was the reaction with a total affect that leaves no doubt that one 'means it,'" as E.H. Erickson observed in his psychoanalytical study of the leading reformist.[42] But as Erikson goes on to say, "we, the heirs of Protestantism, have made convention and pretence out of the very sound of meaning it."[43]

In the 1961 play based on Erikson's influential study, "Martin" (Luther), alone on the stage, screams at the heavens: "Oh Lord, I believe. I do believe. Only help my unbelief!"[44]

The point is this: now that we've all become true believers what is required is the strength of unbelief. That is why I have summoned here the strength of unbelief of the Amerindians and their perspectivist perception, via liberation philosophy and theology, as the worthiest ally. This would be a mode of perception, of reasoning and engaging, itself indissociable from the experience of falling towards and bumping against the "ground" of the totality, without reservation or hope for return (*a fond perdus*, in Adorno's sense). This is also the space and the moment of the totality or the structure's vanishing; a methodological "opening to inexistence," as Badiouians might say.[45]

Boxing theologians!

Let me invoke a Borgesian-Badiouian demarcation of structural vanishing and thinking otherwise as the point at, or the event in which, the totality or structure nearly succeeds in bringing together the orthodox believer and the heretic, as in Borges' short story "The Theologians."[46] Alas, this is the same structural demarcation to be found in Enrique Dussel's philosophical rendition of the term *kairós* as the event of liberation in his reading of the contemporary debates about Paul among philosophers and political theologians (mainly John Milbank, Slavoj Žižek, Alain Badiou and Giorgio Agamben). As said before, this is the mark of liberation theology and philosophy as well as Amerindian-driven anthropologies of permanent decolonization.[47]

It's important to recall that the poor of liberation theology have, in principle, little to do with the poor of neoclassical economics or Rawlsian theories of justice.[48] For liberation theologies and philosophies, "the poor" is a person who dwells in a place (actual or virtual) and not just a representative agent, the "relatively poor" of the space-economy of globalized flows. The poor of liberation philosophy not only dwells, but she's at war. We speak of communities left behind and worst-off generations that constitute our destination, not only as humankind but even beyond humanity, as in the arrangement between humans and nonhumans we call nature. In this respect, the mark of liberation philosophy and the Amerindian anthropology of permanent decolonization is indeed to wage a war of liberation, but not necessarily a war of national liberation. Can we not speak of a war of natural liberation? Let's do so in the sense of the liberation of place in combat with the dispossession of place – permanent decolonization versus permanent "original" accumulation – as the main destination of a history defined not solely in social or biological terms, but also in geological and even cosmological terms, as is the case when we confront 21st-century challenges such as climate change and the Anthropo- or Capitalocene.

As S.J. Pope says apropos of Jon Sobrino's theology, "God is for the poor, *but not for the poor alone*. In other words the universality of divine love is expressed in a special care for those who suffer the most in this world." This radically inclusive love gives rise to hope, which is a civic plebeian virtue, not otherwordly resignation, charitable dignity or justice as fairness. This radically inclusive love or hope principle is what Jon Sobrino, Ignacio Ellacuría and Enrique Dussel identify with *agape*. The latter is construed here as "issuing in mercy"; that is, as letting go (also of the object and nature), as gift and giving, or to dare approach the site of truth as the place where God is absent – an unbearable truth no less,[49] which suffers and struggles with the victims of injustice. This would be a love-driven "fighting universality," to put it in Žižekian terms.[50]

If this is the case, then hope and (attainable) utopia are not a delusion. Hope cannot be understood as being captivated by what does not exist, not yet, but is waited for since it is supposed to come. Rather, in accordance with the Amerindian axiom that "the earth is loaned to us by our children," hope must be conceived of as an experimental device or a forward-looking tool (a technology of speculation or anticipatory realism) that invites us to reconfigure time not only as cyclical but as a continuous loop, as it happens in the mythical stories and philosophies contained in the *Popol Vuh* of the jaguar people or in the myths decoded by Claude Lévi-Strauss.

By contemplating ourselves from the future in order to salvage what has been left by a ruinous past in the present, without reservation or (probabilistic) hedging, come what may, we re-establish the law of reciprocity between the future and the present, as well as the link between our present and the past. The result is to act, in the present, as if the (final) catastrophe that may be projected onto the future had already occurred.[51]

The connection between this speculative device and high scholastic method (the transformation of early modern philosophy impacted upon by Amerindian religious cosmologies and other forms of thinking and philosophizing about contingent claims and the medium of contingency) is best rendered in the commentator's motto, quoted by the priest during his discussion with Joseph K. after the Parable of the Door of the Law in Franz Kafka's *The Trial*: "the right perception of any matter and a misunderstanding of the same matter do not wholly exclude each other."[52] The phrase resonates well with my invocation of Borges and Latin American literature earlier on. It could also resonate with current, challenging understandings of Hegel, but let me stick with the Latin American authors for now. After all, it is well-known that Borges, Gabriel García Márquez and Julio Cortázar were keen readers of Kafka.

Just as in Kafka's story, from the perspective of the priest, the really deluded person in the parable is not the man from the country but the door-keeper himself. More or less could be said of the door-keepers of knowledge and power in the story of the post-16th century commissions told by Lévi-Strauss, which meditates upon their hold on theological and philosophical thinking as well as the social sciences. It is not the indigenous peoples who were duped, but the European door-keepers – the jurists and theologians themselves.

As said before, it's the latter not the former who can be seen as the unbelievers. Historicism and unbelief creep into Christianity when the Christian theologian or his secularized version expresses his faith not only as trust but as certainty and belief in the existence of God just as well. For "he knows that by this very fact it is contestable and contested," as Jean Pouillon puts it.[53] Above all, the Christian or the secularized Christian knows that there are other beliefs because his religion has a history and was constituted against the false gods. But also because this history has not ended since there are still idols and idolatries yet to be eliminated. For instance, capitalism.

And with historicism, as Jon Sobrino observed, "neo-Christendom" also creeps in.[54] This is the imperial "domain of the provision of services and goods."[55] In such a domain, freedom is reduced to "free" market choice aided by probability, and nature enclosed into a reservoir of resources. Once enclosed "out there," as lacking intentionality, and divided "in here" into objects with use and exchange value, not just as commodities, but also as options or derivatives, and thus a subject matter for calculation and "fair" pricing by markets, as the 16th century economical theologian Luis de Molina already implied (probability algorithms applied to markets may be the 21st-century version of a 16th-century debate about pricing and infinitesimals), nature becomes the practico-inert backdrop of the progress of humankind's unique and universal culture.

The former, nature or rather "natural resources," being judged unically as either a curse or a blessing.[56] In the meantime, nature is in fact being pushed to the brink of extinction, together with everything else, us included. What is to be done? The choice is clear: we can either wait in line for the next catastrophe, or else break free from the enclosure together with the poor, future generations and nature itself. The latter is the destination of a philosophy engaged in and pursuing the ends of a war of liberation; this is an action-oriented philosophy inspired by the historical and religious memory of the collective practices and cosmologies of "traditional" and indigenous communities in the Americas (Amerindians, Afro-Americans and Latin Americans, etc.) that for over 500 years have resisted complete colonization by and subordination to industrial rationality and standardization while becoming creolized.

Amerindian religions are creolized religions, not merely hybrid, but rather, intensified forms of cultural mixture: to liberate nature and the dimensionality of times and place from their colonization by the space of (financial/communicative) flows.

Conclusion: liberating nature, or the other side of the story/the other sides of history

This is the perfect point to complete the brief genealogy of religion and the religious ordering of beings that began with the story about post-16th century theological-legal commissions in the Americas providing the dualist matrix (this world/the next one) of our human and practical sciences; and end with a conclusion concerning the ethical-political choice that lies before us.

While European jurists and theologians kept gathering psycho-sociological data on the natives of Haiti, Mexico, central Colombia and southern Brazil, in the neighbouring countries, Amerindians would capture white men and drown them – sometimes in cold water, sometimes in hot water. They would mount guards around the drowned bodies for several weeks, cut them in bits and even taste them in order to find out whether or not they were subject to decay. This is a straight story: the white Europeans who believed in belief trusted to social science; the unbelieving Indians believed in the natural sciences.[57]

The point is not that they all believed in something, that all of them had cultures of their own which should be respected and so on, but the more interesting one is that the Indians saw no reason to accept the Greek-Christian gift of the great divide between nature and culture that ends up rendering the latter into an archipelago of reflexive-identitarian totalities (vgr. tribes, groups, nations, corporations, etc.). And, crucially, of course, their doubt wasn't postmodern![58]

Amerindians refused to accept the discontinuity proposed by the Europeans' encirclement of intentionality as unical and uniquely human, grounding or foundational. The keyword in this argument, crucial to understand the destiny of Christian/modern metaphysics and the progression of belief, is not the term "unique," but the more technical and less familiar term "uncial."

In post-Platonist Christian theology, God is said to be unical in the sense that "he defines and measures all the multitude of beings. For all multitudes being in their own nature indefinite, are bounded through the one."[59] It is precisely in this sense that the humanist matrix of modern humanities and the social sciences, informed by the Greek-Christian gift, the nature/culture divide, is both unique and unical. The unical status granted to the uniquely human reduces natural multiplicity, but also the various cosmological models exploring the multiplication of the multiple, to a case of the one dividing into two. Thus, for instance, our human and social sciences treat these as cases of "human nature" being internally divided into myriad cultural differences – a naturalism ("Western naturalism," according to Viveiros de Castro) supplemented by cultural relativism.[60]

Crucially, mainstream law and anthropology, human rights, economics and so on, declare external human exceptionalism (our alleged "shared values" or "common nature") and set it against internal divisions such as race, gender and especially class. After the critique of philosophical essentialism, we're being told that since there's no a priori grounds for belief or moral engagement then "anything goes" and, consequently, there is no way of discriminating between ethical and unethical actions or to find the right ethical content and wage war for/against it.

That's one reason why, in this time of capitalist globalization and political correctness, the Anthropo-Capitalocene, it has become nearly impossible for the working class to present itself as such (but also for racialized or gendered peoples) not just negatively, but also excessively, and in consequence engage both ethically and politically.

But to say that we cannot discriminate a priori between particular actions does not logically imply that serious moral commitment could not be attached to "less than aprioristically dictated courses of action. To conclude the opposite would be the same as saying that only the particularity of a course of action conceived *as particularity* could be the source of a serious" moral engagement.[61] Workers, racialized and gendered peoples, are conceived of nowadays only qua particularities and their engagements declared or acclaimed as serious or worthy of respect, but nonetheless particular. Such particularities are then opposed to the alleged universality and unity of mankind, and effectively banned as threatening violence against such unity.[62]

This is exactly what the perspectivism found in Amerindian cosmologies and religions without belief denies. Herein lies also the reason for their actuality. The correct analogy is between Amerindian religions without belief, on the one hand, and the mystical discourses of people like Meister Eckhart, John of the Cross and Theresa of Ávila, but also the "God is dead" a-theology and "effective mercy" ethics of the liberationist challenge to Christian theology and Eurocentric philosophy, on the other.

In both cases, it is only insofar as we have the kind of contact with the divinity or the supernatural as an absolute, beyond all particularized content (either the reflexive "I" or the impersonal "It" of the other) that engages the body in all its concrete affectivity (as in Luther's belief, but also in the incomparable dimension of Amerindian inconstancy and the protection privilege of the poor that Herzog speaks of, which have nothing to do with psychological states) that we can give to our particular actions their moral density, seriousness and clarity of object. And only if we sense the effect of the absolute as coming from an utterly empty place, as if we were responding to a "You" spoken by a non-human, can we project onto contingent courses of action a moral depth that, left to themselves, they lack.

This leaves us, indeed, with no a priori rule declaring with assumed authority what are the right incarnating contents or actions. Our belief can only be inconstant, as the Amerindians concluded after their tests.[63]

In both cases, that of Amerindians and that of mystics and liberation philosophers-theologians, if there was an a priori link between the absolute and its incarnating body or content, then the link between the incarnated absolute and its incarnating body would have become a necessary one and the absolute would have lost its excessive dimension. In that case, we would be able to name God or claim mastery in the same way that Christian missionaries claimed discursive mastery over the word and the law, and the king claimed sovereignty over the lands and peoples of the New World in the 16th century.

To state this is not to say that *any content, at any moment* can be an equal candidate for the incarnation of the absolute or the supernatural in nature. Ditto, this would be true only from the point of view of eternity. But just as in Amerindian religions without belief encounters with the supernatural are only contingent and can be lethal for the interlocutor, so, for us, it's clear that serious engagements take place in a terrain that is less than eternity.

That's precisely why for Amerindians, only shamans, strangers, multi-natural beings by office and engagement, not by declaration or manifest destination, are capable of transiting the various perspectives, worm-holing between contexts without possibilities, and balancing the two sides of the ethical equation. Or rather, transforming the entire context.

In their example we can find the answer to our current ethical conundrums. For in construing meaningful ethical lives based on effective mercy and taking responsibility for reality – from mounting inequality to manmade climate change – we must keep open the two sides of the ethical equation.

Yes, an absolute or transcendence that can only be actualized by transforming into something less than itself, and a particularity whose actual engagement is to transcend or shed its own body. And yes, a perpetual alteration. For we humans now occupy the place of enemies through climate change and extractivism, as we hope to be transformed, through death and mercy to the weak, into our enemies/affines, the gods.[64]

Notes

1 Slavoj Žižek, *How to Read Lacan* (London: Granta, 2006) 16–17. See also Eduardo Viveiros de Castro, *Radical Dualism: A Meta-Fantasy on the Square Root of Dual Organizations, or a Savage Homage to Lévi-Strauss* (Amsterdam: Hatje Kantz, 2012).

2 See Frantz Fanon and Raymond Lacaton, "Conducts of Confession in North Africa", previously unpublished transcript of a talk given at the 53rd session of the Congrés des médecins aliénistes et neurologues de France et des pays de langue française, Nice, 5–11 September 1955, now available in *Alienation and Freedom*, ed. by Jean Khalfa & Robert J. C. Young (London: Bloomsbury, 2018) 413–416, for my reference here to

their take on "primitivist" and "modernist" attitudes to madness, spectator's reports and confession in the context of a larger critique of Western political philosophies of social contract that do not acknowledge their contradictory nature in colonial settings. Seen through Fanon's eyes, the problem of the conditions and legal history of confession up to the colonial 20th century echoes the 16th century problem of "inconstancy" among Amerindians. See also, Eduardo Viveiros de Castro, *A inconstância da alma selvagem* (São Paulo: Cosic Naify, 2002) especially 183–266; there's an excerpted English version titled *The Inconstancy of the Indian Soul: The Encounter of Catholics and Cannibals in 16th Century Brazil* (Chicago: Prickly Paradigm Press, 2011).

3 Nicolás Wey Gómez, *The Tropics of Empire: Why Columbus Sailed South to the Indies* (Cambridge, MA: MIT Press, 2008) 409.

4 Silvia Federici, *Calibán y la bruja. Mujeres, cuerpo y acumulación originaria* (Madrid: Traficantes de sueños, 2010/17, 7a. ed.) 294, citing Seymour Phillips, "The Outer World of the Middle Ages", in S. B. Schwartz, *Implicit Understandings: Observing, Reporting, and Reflecting on the Encounters between Europeans and Other Peoples in the Early Modern Era* (Cambridge: Cambridge University Press) 60.

5 Phillips, "The Outer World of the Middle Ages", 61–2.

6 Nigel Biggar, "Don't Feel Guilty about Our Colonial History", in *The Times*, posted 30–11–2017, available at www.thetimes.co.uk/article/don-t-feel-guilty-about-our-colonial-history-ghvstdhmj. For opposing arguments, see Richard Adams, "Oxford University Accused of Backing Apologists of British Colonialism", in *The Guardian*, posted 22–12–2017, available at www.theguardian.com/education/2017/dec/22/oxford-university-accused-of-backing-apologists-of-british-colonialism

7 Federici, *Calibán y la bruja*, 296, for the English version see *Caliban and the Witch* (Autonomedia, 2004) 222.

8 Jane Ana Gordon, "Creolising Political Identity and Social Scientific Method", in *Africa Development*, Vol. 39, No. 1 (Council for the Development of Social Science Research in Africa, 2014, ISSN 0850–3907) 65–80, for my understanding of "creolisation".

9 Álvaro García-Linera, *La potencia plebeya. Acción colectiva e identidades indígenas, obreras y populares en Bolivia* (Clacso/Prometeo Libros, 2013) 16–7, also Jean-Pierre Dupuy, *The Mark of the Sacred* (Stanford: Stanford University Press, 2013) 93, on being "intellectually a Christian" in contrast with "a Christian intellectual".

10 Stephen J. Pope, "On Not 'Abandoning the Historical World to Its Wretchedness': A Prophetic Voice Serving an Incarnational Vision", in *Hope & Solidarity: Jon Sobrino's Challenge to Christian Theology*, ed. by S. J. Pope (New York: Orbis Books, 2008) 44–61, at 53. Pope cites Jon Sobrino, *Where Is God? Earthquakes, Terrorism, Barbarity, and Hope* (New York: Orbis Books, 2006) 136, and *Jesus the Liberator* (New York: Orbis Books, 1993) 17.

11 Jon Sobrino, *Jesus the Liberator*, trans. by P. Burns & F. McDonagh (New York: Orbis Books, 1993) 229.

12 I use the term "Death of God a-theology," in the sense given by Linda Martín-Alcoff and John D. Caputo during a 2005 discussion on Paul's legacy a propos of Žižek's contribution. See John D. Caputo, "Postcards from Paul: Subtraction versus Grafting", in *Paul among the Philosophers*, ed. by John D. Caputo and Linda Martín-Alcoff (Bloomington: Indiana University Press, 2005), 1–23, at 11 and 14. On Amerindian Perspectivism, see Eduardo Viveiros de Castro, *From the Enemy's Point of View: Humanity and Divinity in Amazonia* (Chicago: University of Chicago Press, 1992) 55, and Philippe Descola, *In the Society of Nature: A Native Ecology in Amazonia* (Cambridge: Cambridge University Press, 1996) 93.

13 See on this, Marcel Hénaff, "Living with Others: Reciprocity and Alterity in Lévi-Strauss", in *Rethinking Claude Lévi-Strauss (1908–2009): Yale French Studies*, vol. 123 (New Haven, CT: Yale University Press, 2013) 63–82, at 65. Also Claude Lévi-Strauss, *The Elementary Structures of Kinship* (Boston: Beacon Press, 1969) 51 and 496–497.

14 Serafim Leite, the historian of the Society of Jesus in Brazil, identified "deficiency of will" and "superficial feelings" as the main obstacles to the conversion of the Indians. Relying on a considerable archive of cathechists' records and chronicler's observations,

from Fernández de Oviedo and Antônio Vieira to Gandavo and Capistrano de Abreu, Leite and others speculated this was the reason why Christ had sent his disciple Saint Thomas to preach in Brazil: as punishment. The apostle of doubt was given this most difficult labour of bringing belief to those incapable of believing. See de Castro, *The Inconstancy of the Indian Soul*, 4–5.

15 Among the religious and secular chroniclers, only the Capuchin monk Abbeville seems to have noticed this relation between inconstancy and method: "In truth they are inconstant," he wrote, "if allowing oneself to be guided only by reason may be called inconstancy." But even he mentions reason to the detriment of will, which Amerindians supposedly lack. Ibid. 5. Other examples in Hispanic Latin America include José María de Ripalda and Denis Mesland, Descartes' friend and correspondent.

16 Enrique Dussel, *Philosophy of Liberation*, 158–159, and *Ética de la Liberación*, Editorial Trotta, 1998, 205, 209, 259. Also Eduardo Mendieta's introduction to Enrique Dussel, *Beyond Philosophy: Ethics, History, Marxism, and Liberation Theology*, ed. by E. Mendieta (Rowman & Littlefield Publishers, 2003) 5; and Roberto Goizueta, *Liberation, Method, and Dialogue: Enrique Dussel and North American Theological Discourse* (Scholars Press, 1988) 68–73. See Jon Sobrino, *The Principle of Mercy: Taking the Crucified People from the Cross* (New York: Orbis Books, 1992) 8 ff. and Eduardo Viveiros de Castro, *From the Enemy's Point of View: Humanity and Divinity in Amazonian Perspective* (Chicago: University of Chicago Press, 1992) as well as *Cannibal Metaphysics* (Univocal, 2014).

17 Elsa Dorlin, *Se defendre. Une philosophie de la violence* (Paris: La Découverte, 2017) 19–28.

18 Žižek, *How to Read Lacan*, 14–17, referencing Lévi-Strauss and indirectly evoking the Aristotelian diagram of the three souls. Also, Jean Pouillon, "Remarks on the Verb 'to Believe'", in *A Reader in the Anthropology of Religion*, ed. by M. Lambeck (Blackwell Publishing, 2008) 90–96, at 94.

19 See Roger Haight S. J. "Juan Luis Segundo", in *Empire & The Christian Tradition: New Readings of Classical Theologians*, ed. by Kwok Pui-Lan, Don H. Compier, and Joerg Rieger (Fortress Press, 2007) 439–454, at 443; and Pope, "On Not 'Abandoning the Historical World to Its Wretchedness' 44–61, at 49, for action in solidarity.

20 Bartolomé de Las Casas, *Historia de las Indias* 3.1.2.4, 1994, volumen 3, 1762.

21 Ibid. 3.2.1.9, 1994, volumen 3, 1784–5

22 Ibid. 3.2.1.9, 1994, volumen 3, 1784–86. Also Anthony Pagden, *The Fall of Natural Man: The American Indian and the Origins of Comparative Ethnology* (Cambridge: Cambridge University Press, 1982) 47–50.

23 Juan Ginés de Sepúlveda, *Apologia pro libro de iustis belli*, Argument 4 [1550] 1997, 197. His formulation of the problem of inconstancy is derived from one of the earlier Iberian chroniclers, Gonzalo Fernández de Oviedo's *Historia general y natural de las Indias* 13.6, 1535, xxviir; and, in theological and philosophical terms, from Thomas Aquinas's *In libros Politicorum Aristotelis expositio*, I.1.i.23 and 7.I.5. 968–1127, 1953, 23, 361–364. Another influential source attributed to Aquinas, *De regimine principum*, was largely written by his disciple Ptolemy di Lucca; see his *On the Government of Rulers* 2.1.5. 1997, 105. All sources cited also in Nicolás Wey Gómez, *Tropics of Empire: Why Columbus Sailed South to the Indies* (Cambridge, MA: MIT Press, 2008) 68 n. 181.

24 Claude Lévi-Strauss, *Tristes Tropiques* (London: Penguin Modern Classics, [1955] 2011) 75.

25 de Castro, *The Inconstancy of the Indian Soul* 2–3, citing Antonio Vieira's 1657 *Sermon of the Holy Spirit*.

26 Lévi-Strauss, *Tristes Tropiques*, 75–76, citing the chronicles of Fernández de Oviedo and Ortiz´s 1525 address to the Council of the Indies.

27 Lévi-Strauss, *Tristes Tropiques*, 75, and Raoul Moati, *Derrida/Searle: Deconstruction and Ordinary Language* (New York: Columbia University Press, 2014) on the maintenance of the primacy of conventions in the wake of the "linguistic turn." Also, Christian Lundberg, *Lacan in Public: Psychoanalysis and the Science of Rhetoric* (Tuscaloosa: The University of Alabama Press, 2012), on the debates regarding the nature of the public – the people – and public discourses.

28 Ernesto Laclau, "On the Names of God", in *The Rhetorical Foundations of Society* (London: Verso, 2014) 51.

29 See Alexis Álvarez Nakagawa, *The Cannibal Laws: The Juridical Forms of Conquest and the Emergence of International Law*, submitted and defended for the degree of doctor of philosophy at the School of Law, Birkbeck College, University of London, May 2019, 35–6, for the sources of ideas used in this paragraph.
30 See Lundberg, *Lacan in Public*, 142.
31 See Elvira Roca Barea, *Imperiofobia y leyenda negra. Roma, Rusia, Estados Unidos y el Imperio español* (Barcelona: Siruela, 2016). For a thorough yet balanced criticism of Roca Barea's book, see José Luis Villacañas, *Imperiofilia y el populismo nacional-católico* (Madrid: Lengua de Trapo, 2019).
32 See, Joan Copjec, *Read My Desire: Lacan Against the Historicists* (Boston: The MIT Press, 1996), especially 207 for the social-psychoanalytical on the other as amenable to the exception to the universal predicate. See also, Lundberg, *Lacan in Public*, 147 for the link between Aristotle's conception of stasis and the move from object to objection (*stasis, enstasis*), which I link here to the impossibility of communion and loving integration with the one king or ruler.
33 Ruel, "Christians as Believers, 97–109, at 102.
34 Antonio Vieira, "Sermão do Espírito Santo", in *Sermões* (São Paulo: Editora das Américas [1657] 1957) vol. 5, 205–255.
35 See Fabián Romani Ludueña, *Una comunidad de espectros* (Niño & Dávila editores, 2012), for the enduring impact of anathematic forms of judgment at the heart of our anhropocentric human sciences.
36 Sobrino, *Jesus the Liberator*, 244.
37 Ibid. 269.
38 Joerg Rieger, "Introduction: Watch the Money", in *Liberating the Future: God, Mammon, and Theology*, ed. by Joerg Rieger (Fortress Press, 1998) 10.
39 Ibid.
40 Frederick Herzog, "New Birth of Conscience", in *Liberating the Future: God, Mammon, and Theology*, ed. by Joerg Rieger (Fortress Press, 1998) 147, refering to walking the road that goes from Hispaniola to Wounded Knee "including the auction blocks of black slaves and the abuse of women. Walking this road, we break the spell of fitting Jesus into a universal of religion." See also *Theology Without Foundations: Religious Practice and the Future of Theological Truth*, ed. by Stanley Hauerwas, Nancy Murphy and Mark Nation.
41 Jean Pouillon, "Remarks on the Verb 'to Believe'", in *A Reader in the Anthropology of Religion*, ed. by M. Lambeck (Blackwell Publishing, 2008) 90–96, at 93.
42 Erik Homburgen Erikson, *Young Man Luther* (London: Faber & Faber, 1959) 203.
43 Ibid.
44 John Osborne, *Luther* (Faber & Faber, 1961), the two final scenes. Notice the homology in the final scenes of the play between Martin's cry and Jesus's cry on the cross, "Father, why have you forsaken me!" See also Slavoj Žižek, On Belief, 106–151.
45 See Alain Badiou, *Logic of Worlds*, Continuum, 2009, and 'Politics: A Non-Expressive Dialectics' in *Philosophy for Militants*, 75, for my use of "inexistence" and "vanishing."
46 Jorge Luis Borges, "The Theologians", included in *Labyrinths*, Penguin, 1970, 150, also "Del rigor en la ciencia", in *El hacedor* and "On Exactitude in Science", in *A Universal History of Infamy*, Penguin, 1975. Here again I follow E, Kaufinan, *Do Dual Structures Exist?* 93. See also G. García Márquez, "Posibilidades de la antropofagia", in *Anthropofagia Hoje?*, ed. by J. Ruffinelli and J. C. De castro Rocha (Sao Paulo: Editorial Realizacoes, 2011) 47.
47 Enrique Dussel, *Kairós: El acontecimiento liberador en Pablo de Tarso*, mauscript on file with the autor, 2009. Also available at www.enriquedussel.com/txt/II-CAP-4-31.pdf
48 In such conceptualisations, the law of reciprocity is circumvented by linear narratives of historical time. In such narratives, current beneficiaries of past injustice are said to owe nothing in return to those from whom they received so much, and, conversely, those who come later receive something from preceding generations but can give nothing back. Worse still, since in such narratives progress dictates that "developed" societies and future generations must be wiser and better, but moral duty rests with the weaker

societies, the poor, and current generations, it is only the latter who are said to have a moral burden. This adds to the historical injuries inflicted upon less-developed nations, the poor and worst-off present/future generations an insult: you're not really "poor," you just happen to occupy a relatively low rank in a hierarchy of poverty; as such, you can aspire to advance, and you can do so since only you're responsible for your own condition, only you can give! Such theories of justice, and the economic theologies and teleologies that underpin them, constitute an option for the poor *alone*. As such, they´re bad explanations of engaged belief and even worse blueprints for political action. Given the theoretical insult that is added nowadays to historical injustices, is there really any surprise in the fact that weaker societies, the poor and worst-off present/future generations cannot help but feel resentment (Adam Smith's envy/sympathy) and give in to violence? See jaen-Pierre Dupuy, *The Mark of the Sacred*, Chapter 5; and Marcel Hénaff, *The Price of Truth: Gift, Money and Philosophy* (Stanford University Press, 2010).

49 Pope, "On Not 'Abandoning the Historical World to Its Wretchedness', 48, for the prophet/priest dialectic, and 57 for agape as issuing, releasing or liberating in mercy. Also Jon Sobrino, *Jesus the Liberator* (New York: Orbis Books, 1993) 34.

50 Pope, "On Not 'Abandoning the Historical World to Its Wretchedness'", 51.

51 Popol Vuh, *Relato Maya del origen del mundo y de la vida* (Madrid: Trotta, 2008); and Francisco J. Varela, "The Dance: Subjectivity-Objectivity", available at www.youtube.com, for a scientific distant relative of perspectivism. Also Oscar Guardiola-Rivera, *Being against the World* (London: Routledge and Birkbeck Law Press, 2009); Eugene Thacker, *In the Dust of This Planet* (London: Zero Books, 2011); Dupuy, *The Mark of the Sacred*, 62–63; and Evan Calder Williams, *Uneven and Combined Apocalypse* (London: Zero Books, 2011), for catastrophism and contemplating disaster from the future.

52 Franz Kafka, *The Trial* (Penguin Books, 1985) 238, cited by Slavoj Žižek, *On Belief* (London: Routledge, 2001) 110.

53 After all, the Christian cannot avoid expressing his faith not only as trust in God (confiance en, in French, confianza en, in Spanish), but also as belief in (croyance à, in French, creencia en, in Spanish) his existence and belief that (croyance que, in French, creer que, in Spanish) "God possesses such and such attributes, that the world was created, and so forth." See Pouillon, *Remarks on the Verb 'to Believe'*, 94.

54 Jon Sobrino, *Where Is God? Earthquakes, Terrorism, Barbarity, and Hope* (New York: Orbis Books, 2006) 136, and *Jesus the Liberator* (New York: Orbis Books, 1993), 17.

55 Žižek, *On Belief*, 113, citing both Hannah Arendt and Alain Badiou.

56 This is in accordance to the anathematic form of Christian judgement.

57 Lévi-Strauss, *Tristes Tropiques*, 76.

58 This is a reference to Bellah's paper "The Historical Background of Unbelief", included in *The Culture of Unbelief. Proceedings from the First International Symposium on Belief*, Rome, 22–27 March, 1969: also to the continous progression from belief-as-batch to belief-as-inward-experience and from then onto belief-as-shared values (albeit different) evident in *Christian Believing, Doctrine Commission of the Church of England*, SPCK, 1976, both cited by Malcolm Ruel, *Christians as Believers*, 105. There was no case of "diffuse belief" among Amerindians or a comparable one central to them in the same way belief is for Christians.

59 See Proclus, *The Six Books of Proclus, the Platonic Successor, on the Theology of Plato*, volume II, Proposition CXVII, 380–384, trans. by T. Taylor, 1816, digitized by The University of Michigan, April 2008.

60 Similarly, infinity becomes the finite (i.e., the uniquely human activity of the intellect drawing the line between the sensible and the intelligible, as well as projecting upon the globe, in space and time, a horizontal line dividing what is within sight from what lies beyond).

61 Laclau, *The Rhetorical Foundations of Society*, 37–51, at 50–51. Here and in what follows I take the direction explored by Laclau's investigation of mystical discourses on the names of God.

62 Ibid.

63 It's lack of a priori rules explains the experiments of Amerindians. In testing whether or not Europeans had bodies, whether they would rot or resurrect, they were not only trying to falsify the promise of advent. They were not merely testing the facts, but testing the ways in which we can establish what are the facts. It's in this respect that the Amerindians (and the mystics and liberation theologians) appear as the champions of facticity, insofar as they deny the existence of an aprioristic logic linking the encounter with the absolute to particular actions or contents. Amerindians see such encounters as purely contingent. For instance, suddenly finding out in the middle of the jungle that the other is human, that "It" is human. All of a sudden being called "You" by the non-human subjectivities and spirits that populate nature and the supernatural, thus shedding one's particular body to transform into another, thereby acquiring a perspective that is less than absolute and yet not merely that of a particular body. See Eduardo Viveiros de Castro, "Deixis and Amerindian Perspectivism", in *The Journal of the Royal Anthropological Institute*, vol. 4, no. 3, September 1998, 469–488, at 483, also included in *A Reader in the Anthropology of Religion*, 280–297.

64 Ibid. See also de Castro, *The Inconstancy of the Indian Soul*, 103; and "Deixis and Amerindian Perspectivism", 469–488, at 483 for borrowings and paraphrases; and for context, Bruno Latour, "Perspectivism: Type or Bomb?", in *Anthropology Today*, vol. 25, no. 2, April 2009, London & Oxford: Blackwell, 1, referring to the work of E Viveiros de Castro; and T Stolze Lima, "The Two and Its Many: Reflections on Perspectivism in a Tuna Cosmology", in *Ethnos*, vol. 64, no. 1, 1999, 107–131. For humans occupying the place of enemies in relation to manmade climate change and extractivism, and moving beyond anthropocentrism in history and the humanities, see Dipesh Chakravarty, "The Climate of History", in *Critical Inquiry*, 35, The University of Chicago, Winter 2009, 197 ff.

Bibliography

Adams, Richard. "Oxford University Accused of Backing Apologists of British Colonialism", in *The Guardian*, published 22-12-2017, available at www.theguardian.com/education/2017/dec/22/oxford-university-accused-of-backing-apologists-of-british-colonialism

Álvarez Nakagawa, Alexis. *The Cannibal Laws: The Juridical Forms of Conquest and the Emergence of International Law*, submitted and defended for the degree of Doctor of Philosophy at the School of Law, Birkbeck College, University of London, May 2019.

Aquinas, Thomas. *In octo libros Politicorum Aristotelis expositio*, I.1.i.23 and 7.I.5. 968–1127, Rome: Marietti, 1953 and 1966.

Badiou, Alain. "Politics: A Non-Expressive Dialectics", in *Philosophy for Militants*, London: Verso, 2012.

Badiou, Alain. *Logic of Worlds*, London: Continuum, 2009.

Biggar, Nigel. "Don't Feel Guilty about Our Colonial History", in *The Times*, posted 30–11–2017, available at www.thetimes.co.uk/article/don-t-feel-guilty-about-our-colonial-history-ghvstdhmj.

Borges, Jorge Luis. "The Theologians", in *Labyrinths*, London: Penguin, 1970, 150.

Borges, Jorge Luis. "On Exactitude in Science", in *A Universal History of Infamy*, London: Penguin, 1975.

Caputo, John D. "Postcards from Paul: Subtraction versus Grafting", in *Paul Among the Philosophers*, ed. by John D. Caputo and Linda Martín-Alcoff, Bloomington: Indiana University Press, 2005, 1–12.

Calder Williams, Evan. *Uneven and Combined Apocalypse*, London: Zero Books, 2011.

Chakravarty, Dipesh. "The Climate of History", in *Critical Inquiry*, 35, Chicago: The University of Chicago, Winter 2009, 197 ff.

Copjec, Joan. *Read My Desire: Lacan Against the Historicists*, Boston: The MIT Press, 1996.

Descola, Philippe. *In the Society of Nature: A Native Ecology in Amazonia*. Cambridge: Cambridge University Press, 1996.

Dorlin, Elsa. *Se defendre. Une philosophie de la violence*. Paris: La Découverte, 2017.

Dupuy, Jean-Pierre. *The Mark of the Sacred*. Stanford: Stanford University Press, 2013.

Dussel, Enrique. *Kairós: El acontecimiento liberador en Pablo de Tarso*, mauscript on file with the autor, 2009. Also available at www.enriquedussel.com/txt/II-CAP-4-31.pdf

Dussel, Enrique. *Ética de la Liberación*, Madrid: Editorial Trotta,1998.

Erikson, E. H. *Young Man Luther*, London: Faber & Faber, 1959.

Fanon, Frantz and Raymond Lacaton, "Conducts of Confession in North Africa", in *Alienation and Freedom*, ed. by Jean Khalfa and Robert J. C. Young, London: Bloomsbury, 2018.

Federici, Silvia. *Calibán y la bruja. Mujeres, cuerpo y acumulación originaria*. Madrid: Traficantes de sueños, 2010/17, 7a. English version titled *Caliban and the Witch*, Minneapolis: Autonomedia, 2004.

Fernández de Oviedo, Gonzalo. *Historia general y natural de las Indias* 13.6, 1535, xxviir, Madrid: Imprenta de la Real Academia de Historia, 1852.

García-Linera, Álvaro. *La potencia plebeya. Acción colectiva e identidades indígenas, obreras y populares en Bolivia*. Clacso/Prometeo Libros, 2013.

García Márquez, Gabriel. 'Posibilidades de la antropofagia', in *Anthropofagia Hoje?*, ed. by J. Ruffinelli and J. C. De Castro Rocha, Sao Paulo: Editorial Realizacoes, 2011, 47.

Ginés de Sepúlveda, Juan. *Democrates secundus, sive Apologia pro libro de iustis belli*, Argument 4, ed. by A. Corolet Lletget and A. Moreno H., in *Obras Completas* III, 38–134, Pozoblanco: Ayuntamiento de Pozoblanco [1550] 1997.

Goizueta, Roberto. *Liberation, Method, and Dialogue: Enrique Dussel and North American Theological Discourse*, Scholars Press, 1988.

Gordon, Jane Ana. "Creolising Political Identity and Social Scientific Method", in *Africa Development*, Volume XXXIX, No. 1 (Council for the Development of Social Science Research in Africa, 2014) 65–80.

Guardiola-Rivera, Oscar. *Being Against the World*, London: Routledge & Birkbeck Law Press, 2009.

Haight S. J., Roger "Juan Luis Segundo", in *Empire & The Christian Tradition. New Readings of Classical Theologians*, ed. by Kwok Pui-Lan, Don H. Compier, and Joerg Rieger. Fortress Press, 2007, 439–454.

Hénaff, Marcel. "Living with Others: Reciprocity and Alterity in Lévi-Strauss", in *Rethinking Claude Lévi-Strauss (1908–2009). Yale French Studies*, volume 123, New Haven, CT: Yale University Press, 2013, 63–82.

Hénaff, Marcel. *The Price of Truth. Gift, Money and Philosophy*, Stanford: Stanford University Press, 2010.

Herzog, Frederick. "New Birth of Conscience", in *Liberating the Future: God, Mammon, and Theology*, ed. by Joerg Rieger, Fortress Press, 1998, 147–159.

Kafka, Franz. *The Trial*, Penguin Books, 1985.

Laclau, Ernesto. "On the Names of God", in *The Rhetorical Foundations of Society*, London: Verso, 2014.

Las Casas, Bartolomé de. *Historia de las Indias* 3.1.2.4, volumen 3, Madrid: Alianza Editorial, [1762] 1994.

Latour, Bruno. "Perspectivism: Type or Bomb?", in *Anthropology Today*, vol. 25, no. 2, April 2009, London and Oxford: Blackwell, 1.

Lévi-Strauss, Claude. *Tristes Tropiques*, London: Penguin Modern Classics, [1955] 2011.

Lévi-Strauss, Claude *The Elementary Structures of Kinship*, Boston: Beacon Press, 1969.

Lucca, Ptolemy di. *On the Government of Rulers* 2.1.5. University of Pennsylvania Press, 1997.

Lundberg, Christian. *Lacan in Public. Psychoanalysis and the Science of Rhetoric*, Tuscaloosa: University of Alabama Press, 2012.

Mendieta, Eduardo. "Introduction", in Enrique Dussel's *Beyond Philosophy: Ethics, History, Marxism, and Liberation Theology*, ed. by E. Mendieta, Lanham, MD: Rowman and Littlefield Publishers, 2003.

Moati, Raoul. *Derrida/Searle. Deconstruction and Ordinary Language*, New York: Columbia University Press, 2014.

Pagden, Anthony. *The Fall of Natural Man: The American Indian and the Origins of Comparative Ethnology*, Cambridge: Cambridge University Press, 1982.

Phillips, Seymour. "The Outer World of the Middle Ages", in S. B. Schwartz, *Implicit Understandings: Observing, Reporting, and Reflecting on the Encounters Between Europeans and Other Peoples in the Early Modern Era*. Cambridge: Cambridge University Press, 1995.

Pope, Stephen J. "On Not 'Abandoning the Historical World to Its Wretchedness'. A Prophetic Voice Serving an Incarnational Vision", in *Hope & Solidarity. Jon Sobrino's Challenge to Christian Theology*, ed. by S. J. Pope, New York: Orbis Books, 2008.

Pouillon, Jean. "Remarks on the Verb 'to Believe'", in *A Reader in the Anthropology of Religion*, ed. by M. Lambeck, Oxford: Blackwell Publishing, 2008, 90–96.

Proclus, *The Six Books of Proclus, the Platonic Successor, on the Theology of Plato*, volume II, Proposition CXVII, 380–384, trans. by T. Taylor, 1816, digitized and published by The University of Michigan, April 2008.

Rieger, Joerg. "Introduction: Watch the Money", in *Liberating the Future: God, Mammon, and Theology*, ed. by Joerg Rieger, Fortress Press, 1998.

Roca Barea, Elvira. *Imperiofobia y leyenda negra. Roma, Rusia, Estados Unidos y el Imperio español*, Barcelona: Siruela, 2016.

Romani Ludueña, Fabián. *Una comunidad de espectros*, Niño and Dávila editores, 2012.

Ruel, Malcolm. "Christians as Believers", in *A Reader in the Anthropology of Religion*, ed. by M. Lambeck, Oxford: Blackwell Publishing, 2008, 97–109.

Sobrino, Jon. *Where Is God? Earthquakes, terrorism, Barbarity, and Hope*, New York: Orbis Books, 2006.

Sobrino, Jon. *Jesus the Liberator*, trans. by P. Burns and F. McDonagh, New York: Orbis Books, 1993.

Sobrino, Jon. *The Principle of Mercy: Taking the Crucified People from the Cross*, New York: Orbis Books, 1992.

Stolze Lima, Tania. "The Two and Its Many: Reflections on Perspectivism in a Tuna Cosmology", in *Ethnos*, vol. 64, no. 1, 1999, 107–131.

Thacker, Eugene. *In the Dust of This Planet*, London: Zero Books, 2011.

Vieira, Antonio. "Sermão do Espírito Santo", in *Sermões*. São Paulo: Editora das Américas [1657] 1957, vol. 5, 205–55.

Villacañas, José Luis. *Imperiofilia y el populismo nacional-católico*, Madrid: Lengua de Trapo, 2019.

Viveiros de Castro, Eduardo. *Cannibal Metaphysics*, Minneapolis: Univocal, 2014.

Viveiros de Castro, Eduardo. *A inconstância da alma selvage*, São Paulo: Cosic Naify, 2002. English version titled *The Inconstancy of the Indian Soul. The Encounter of Catholics and Cannibals in 16th Century Brazil*, Chicago: Prickly Paradigm Press, 2011.

Viveiros de Castro, Eduardo. *Radical Dualism: A Meta-fantasy on the Square Root of Dual organizations, or a Savage Homage to Lévi-Strauss*, Amsterdam: Hatje Kantz, 2012.

Viveiros de Castro, Eduardo. "Deixis and Amerindian Perspectivism", in *The Journal of the Royal Anthropological Institute*, vol. 4, no. 3, September 1998, 469–488, at 483, also

included in *A Reader in the Anthropology of Religion*, ed. by M. Lambeck, Oxford: Blackwell Publishing, 2008, 280–297.
Viveiros de Castro, Eduardo. *From the Enemy's Point of View. Humanity and Divinity in Amazonia*. Chicago: University of Chicago Press, 1992.
Vuh, Popol. *Relato Maya del origen del mundo y de la vida*, Madrid: Editorial Trotta, 2008.
Wey Gómez, Nicolás. *The Tropics of Empire. Why Columbus Sailed South to the Indies*. Cambridge, MA: MIT Press, 2008.
Žižek, Slavoj. *How to Read Lacan*, London: Granta, 2006.
Žižek, Slavoj. *On Belief*, London: Routledge, 2001.

Filmography

Varela, Francisco J. "The Dance: Subjectivity-Objectivity", available at www.youtube.com

8

INDIGENOUS PEOPLES AND THE ENVIRONMENT

Views from Brazil

Seth Garfield

> We, the Kayabi, Apiaká, and Munduruku indigenous peoples from the lower Teles Pires River and Rikbaktsa from the lower Juruena River, gathered in Teles Pires village between the 21st and 24th of April 2015, reaffirm our alliance and unity in defense of the rivers Teles Pires, Juruena, and Tapajós. The government builds dams with hastily-prepared and incomplete environmental studies, without seeking to understand the consequences of the destruction of nature for our lives, authorizing the operation of dams without providing a response to indigenous people about how they will continue their lives without fish, without water, without hunting. It tries to hide their negative impacts on our lives, our rivers and our territories. The government doesn't bring information that we understand, in our villages and in our languages, nor does it offer alternatives for our physical and cultural survival. . . . The Federal government has not respected our right to free, prior, and information consultation and consent, guaranteed by the Federal Constitution and by International Labour Organization Convention 169, prior to making political decisions concerning the construction of dams on the Teles Pires River.
>
> ('Manifesto of the Alliance' 2016)

Amidst the dramatic transformation of the natural environment of the Americas in the half millennium following European conquest, this manifesto by besieged native Amazonian communities stands out as much for its historical conventionalism as its innovation. If we take as a truism, even a redundancy, Alfred Crosby's notion of "ecological imperialism" – namely, that Europe's global expansion was predicated upon and enabled by the concerted assault on indigenous peoples' ecosystems – then the native Amazonians' lament of fish, water and hunting game sacrificed on the developmental altar of the Brazilian state reveals the enduring disposability of indigenous lifeways (Crosby 1986). If we apprehend indigenous peoples' shared postcolonial traumas of dispossession as shaded merely by differing patterns of capital accumulation, then the presentiments of the Tapajós river basin's natives conform to the variations of an historical theme (Ribeiro 1970). And if we take as a commonplace that indigenous peoples of Latin America have long used laws, alongside other armatures of conquest assembled by colonizers in their "civilizing" missions, to defend their access to land and natural resources, then the indigenous complainants follow a long line of native petitioners and litigants who have brandished their rights as subjects and citizens to parry the systematic patterns of legal abuse, non-enforcement and impunity that endanger their "physical and cultural survival."

Yet there is much in the manifesto, particularly the indigenes' confrontation with Goliath-like hydroelectric dams and their strategic deployment of a scientific-environmental counter narrative that reflects newer chapters in the centuries-old history of native peoples' struggles over resources and rights in South America. Hydroelectric dams, satisfying growing demand for electrical power in Brazil, exemplify (and extend) the vertiginous social and environmental transformations of the 20th and 21st centuries whose unequal impacts are decried globally by their most vulnerable victims. In 1900, for example, there were approximately 600 large dams worldwide; by 2000, there were approximately 45,000 (Khagram 2004: 1–10). Similarly, moralistic appeals might have served native peoples in rebuking colonial dominators' not-so-Christian behaviour or in reprimanding their newly minted postcolonial oppressors, but the Amazonian natives' self-ascribed role as conservators of "biodiversity, which is considered World Heritage, and for which the non-indigenous people have proved to be incapable of governing," evinces the reframing of indigenous struggle and identity in global eco-political terms. Although the constitutional rights of Brazil's indigenous populations stem from their status as first peoples, native peoples have deployed ecological claims to establish alliance if not always common ground – with environmentalist sceptics of capitalist excess, technocratic supremacy and natural resource sustainability (Conklin and Graham 1995; Dove et al. 2003: 19–46).

This essay historicizes contemporary conflicts involving the indigenous peoples of Brazil and their relationships to the natural environment. Native Brazilians' ties to the natural environment, whether in the productive, symbolic or affective realms, can be understood as "the collective, historically contingent identities, ideologies, and environmental knowledge systems" that indigenous groups establish to maintain territory through interactions grounded in relationships of power (Little 2001: 5; Wright, Kapfhammer, and Wiik 2012). With a population of 896,900, native peoples represent less than 0.5 percent of Brazil's national population. As defined by the Brazilian Indian Statute of 1973, an indigenous person is "any individual of pre-Columbian origin and descent who self-identifies or is identified as belonging to an ethnic group whose cultural characteristics differentiate him [or her] from national society." Brazil's Constitution of 1988 also guarantees that indigenous peoples shall have "their social organisation, customs, languages, creeds and traditions recognized, as well as their original rights to the lands they traditionally occupy." As of 2010, there were over 500 indigenous areas demarcated in Brazil, comprising 12.5 percent of national territory and containing a population of 517,400 (Portal Brasil 2015).

To be sure, Brazil's indigenous peoples are marked by remarkable diversity. According to the national census of 2010, there are more than 300 indigenous groups in Brazil, speaking over 270 languages. One third of indigenous groups in Brazil number fewer than 200 people each (Survival International 2000: 58). Historical timelines of adaptation to dominant society vary markedly – in 2015, alongside the descendants of the first peoples encountered by the 16th-century Portuguese, were 26 indigenous groups that remained isolated from dominant society, and more than 50 others awaiting official survey. Regional forms of capital accumulation and infrastructural development – whether mining, ranching, extractivism, road-building or hydroelectric dams – have transformed indigenous peoples' cultural ecology in distinctive ways. Varied histories of proselytising by Catholic and evangelical Protestant missionaries, as well as adherence to traditional beliefs, serve to cross-cut indigenous peoples' cosmovisions.

Indigenous peoples' relationships to the natural environment also range significantly. Present throughout Brazil – with the largest concentration (324,800) in the Northern (Amazon) region – indigenous peoples' homelands stretch from tropical forests to savannahs to grasslands to semi-arid backlands to urban areas, where 36.2 percent of the native population now resides (Portal Brasil 2015). In general, indigenous groups that are more remote from commercial markets tend to maintain traditional forms of environmental management linked to kin-based and mythological understandings of nature, whereas those groups restricted to small reservations, closer to urban centres and subject to extreme privation tend to place heavier strain on natural resources, suffer reduced subsistence and rely more heavily on market exchange (Lima and Pozzobon 2005: 50–51). Even among native Amazonians more remote from commercial markets, scholars have noted a tendency to view the environment as an "association of landscapes," each possessing a different character and essence, rather than as a monolithic entity (Balée 2003: 277).

Nonetheless, in mapping indigeneity in Brazil according to the coordinates of capitalist development, historical demography and global ecological change, we can sustain fundamental metanarratives. Capitalism, in its very origins, entailed a revolutionary transformation in human relations with the (non-human) natural environment grounded in commodity production, territorial enclosure, and the immiseration of peasantries and indigenous populations (Moore 2015). The settlement of the Americas, advanced through the implantation of European-style agriculture and the slave-plantation complex, involved a systematic assault on Native American ecologies and attendant lifeways. Epidemic disease, genocide, enslavement, forced resettlement and acculturation decimated Brazil's pre-Columbian population from an estimated five million to one million by the early 20th century, and reduced native peoples from a majority of Amazonia's population to a minority (Ribeiro 1970; Moreira Neto 1988; United Nations 2016). And between 1900 and 1957, Brazil's indigenous population plunged to less than 200,000, with over 80 indigenous groups destroyed by disease and violence resulting from contact and conflict with ranchers, miners, extractivists and railroad construction crews (Davis 1977: 5).

An understanding of contemporary environmental conflicts engulfing Brazil's indigenous population also requires attentiveness to more recent demographic trends. Brazil's native population expanded by 205 percent between 1990 and 2010, from approximately 294,000 to slightly under 900,000, due to acquired immunity to epidemic diseases, the expansion of indigenous health care (although native peoples remain woefully underserved), the uptick in demarcation of indigenous territories since the 1990s (affording communities greater physical security and psychological well-being), and a greater tendency for individuals and communities with longer histories of integration into dominant society to self-identify as indigenous (Inter-American Commission on Human Rights 1997; Oliveira Filho 1999; Warren 2001; Portal Brasil 2015). This rapid demographic growth, amidst cramped reservations and adjacent ecosystems denuded by developers, has increased strain on natural resources and inflamed intra- and interethnic conflict. The indigenous population also remains plagued by high rates of malnutrition, alcoholism and poverty. In 2000, for example, Native Brazilians had a life expectancy of 42 years compared to 67 for the rest of the country (Jones 2000).

Moreover, if we employ a composite definition of "nature" as an ecosystem with energy flow and nutrient recycling; a physical landscape whose spatial context and natural resources offer living species varied potentials for use, transformation or conservation; and a cultural construction shaped by and constitutive of social practices and values and ontological notions, the lines of battle for indigenous people in Brazil are also often clear (Sponsel 2011: 40–41). As Rob Nixon notes more globally, indigenous, or "vernacular," landscapes, while neither monolithic nor uncontested, are shaped by the maps that local communities have devised (and revised) over generations to ensure access to ecological resources, to maintain affective and spiritual bonds, and to inscribe and honour historical memory. In "official" landscapes, whether those charted by government planners, corporate investors or environmental organizations, such intricate historical/social/cosmological ties are discounted by high-modernist developmental projects, short-term extractivist agendas and conservationist goals (Nixon 2011: 17).

Since indigenous peoples in Brazil have modified natural ecosystems for millennia – transforming forests, manipulating and sustaining biodiversity, domesticating crops, marketing goods, redistributing species and channelling rivers – present-day conflicts are not best understood as instances of "pristine" nature violated by developers, rescued by conservationists or even unanimously beheld by local communities beset by poverty and need (Argawal and Gibson 1999). All societies transform the natural world, albeit in accord with localized and culturally specific forms of environmental knowledge and practice that are better understood as makers and markers of particularistic worldviews rather than objective truths (Diegues 2004; Descola 1994; Descola 2013). Nor should such disputes between "the West" and "the rest" serve to cast the interpenetration of "nature" and "culture" as exceptional to the mental geographies of traditional peoples rather than variations of a general pattern of world – making that modern actors (and their critics' ideological inversions) have occluded (Latour 1993; Argawal 1995; Descola 2013). Rather, these contested landscapes reflect how longstanding competitions over ecological resources, knowledge and imaginaries have evolved across the more recent histories of statist and neo-liberal forms of capitalism, the unprecedented scale of ecological destruction characteristic of industrial consumerism and the concomitant rise of the global environmental movement.

Fuelled by voracious consumer demand in the United States, Europe and China, and the more affluent regions of Brazil, a recent wave of land grabs, resource plunder, road-building and large dam projects over the last decades has left a trail of devastation in indigenous communities (Nixon 2011: 22). Colonial policies prioritizing the coercion of native labour for mercantilist gain and settler benefit have been sidelined by the premiums placed on indigenous lands and resources. Although the demarcation of reservations has increased over the last two decades in response to indigenous political mobilization, international pressure and capitalist developmental initiatives to regularize land titles in the Brazilian countryside, many native territories remain invaded, despoiled and contested. Violence has surged. In 2014, 138 indigenous people were murdered in Brazil, frequently resulting from conflicts linked to land demarcation and resource appropriation (Conselho Indigenista Missionário 2015). The following year, Greenpeace named Brazil "the most dangerous country for environmental activists," accounting for 50 of the 185 murders worldwide linked to conflicts over agribusiness expansion and the construction of hydroelectric plants. In fact, indigenous peoples worldwide represented a disproportionate 40 percent of those slain (Greenpeace 2016a).

In her visit to Brazil in March 2016, the U.N. Special Rapporteur on the Rights of Indigenous Peoples found a "worrying regression in the protection of indigenous peoples' rights," notwithstanding the country's "exemplary constitutional provisions" (United Nations 2016). The discussion that follows focuses on discrete historical contexts and contests impacting Native Brazilians' ties to the natural environment that are fundamental to understand their social and cultural reproduction and the contemporary challenges facing their communities.

Indigenous peoples and the myth of pristine environments

For state developers and private investors, native forms of land use in Brazil are deemed unscientific, unproductive, unchanging, unimportant or merely unimagined (Ramos 1998). Like other manifestations of environmental racism, such canards are convenient to legitimate marginalization and dispossession (Nixon 2011). They are also historically inaccurate. The paleoecological, ethnolinguistic and historical record point to millennia of indigenous modification of Brazil's landscapes reflective of native social organization, cultural ecology and cosmological belief.

Numerous scholars have now debunked the myth of pristine pre-Columbian landscapes, including the multiple ways in which Native Brazilians transformed forests prior to the arrival of the Portuguese in 1500 (Denevan 1992; Denevan 2003; Mann 2005; Balée 2013). Landscape domestication may be defined as a "conscious process by which human manipulation of the landscape results in changes in landscape ecology and in the demographics of its plant and animal populations" for the benefit of humans. Plant domestication is best understood as

> a co-evolutionary process by which human selection of the phenotypes of promoted, managed or cultivated plant populations results in changes in the population's genotypes that make them more useful to humans and better adapted to human intervention in the landscape.

At the time of conquest, over 138 crops plant species were cultivated, managed or promoted by Amazonian peoples, including manioc, papaya, cashew, guava, cocoyam, guaraná, pineapple and chili peppers; they also cultivated and managed many tree species, including Brazil nuts, peach palms and soursop (Clement 1999a: 190–192). At least 86 native plant species were domesticated by Amazonian peoples to varying degrees, creating new agricultural landscapes by spreading crops around the basin. It is estimated that 90 percent of the crop genetic diversity in Amazonia at the time of contact was lost over the next centuries (Clement 2006: 38).

Indigenous peoples of Brazil have used fire for millennia to modify both forest and savannah environments, whether to hunt game, encourage the regeneration of select vegetation, remove excess leaves and branches, enrich soil nutrients (particularly calcium, magnesium and potassium) or prevent invasion of savannah by forest species (Allen 1998: 56). In the Atlantic forests that blanketed some 1.3 million square miles of Brazil's eastern coastal regions prior to the arrival of the Portuguese, the Tupi-Guarani people had long used fire to clear small areas for subsistence cultivation. The regeneration of the forest was ensured by the reduced size of the openings and the lengthy fallow periods that typically lasted between 20 and 40 years (Dean 1995: 27; Pádua 2010: 139–140).

During the early colonial period, sustained and successive deforestation by Portuguese settlers paved the way for sugar monocrop expansion and the provisioning of food and fuel for the 18th-century mining centres. In 1711, the Jesuit André João Antonil noted of the indiscriminate destruction of tropical forests: "once the choice of the best land for the sugar cane had been made, it is cleared, burnt, cleaned and everything that can serve as a hindrance must be removed from there." The Luso-Brazilian assault on the Atlantic forests – stemming from seemingly inexhaustible supplies of choice soils and wood, strategies to legitimize land claims, and the demonization of tropical landscapes and populations – is also evident in Antonil's wish that Brazil's sugar mills "with the immensity of forests that it has could enjoy, as it had for years, and would in the future, as many furnaces as existed now" (Pádua 2010: 137, 143).

The Tupi-Guarani peoples of the Atlantic forests witnessed a concomitant demographic collapse. A pre-Columbian population estimated at one million had dropped to approximately 50,000 by 1600. Although many native peoples had fled to the backlands to escape the coastal redoubts of European settlement, the depopulation owed largely to the socio-environmental consequences of Iberian colonialism: the decimation of native peoples lacking immunological defense to Old World pathogens, the systematic enslavement and dispossession of native communities, and the destruction of forests and ensuing degradation of soils that served to undermine traditional subsistence strategies and compound the effects of epidemic disease (Pádua 2010: 132, 140).

The 19th-century coffee boom in southeastern Brazil augmented the devastation of Atlantic forests and the depletion of soil fertility from monocrop production (Baer 2001: 386). By the early 20th century, the Atlantic tropical forest had been ravaged by total deforestation rates estimated at between eight and 28 percent; and by the end of the century, the original forest had been reduced by 92 percent (Pádua 2010: 130–133). Indigenous lifeways tied to the historical ecologies of Atlantic tropical forests would be forever diminished or destroyed.

In the Amazon, which supported large indigenous populations prior to the European conquest, evidence of human modification of forests and landscapes abounds. The swidden-fallow sequence, the most common indigenous practice in the Amazon today, entails the combination of cultivated landscape (swidden) with high yields for several years prior to the decrease in soil fertility, followed by management of weeds and transplanted shrubs and trees until a managed secondary forest (the fallow) ensues (Moran 1993; Allen 1998: 56–57; Clement 2006: 35). Fallow forests also typically contain nutrient-richer soils, known as *terra preta do índio*, a man-made black soil of relatively high pH resulting from long-term accumulated refuse in middens, including ash and charcoal from domestic fires. Dark-earth soils often contain large concentrations of ceramics, bones and kitchen wastes, pointing to their origins in former settlement sites created between 1000 B.C. and A.D. 1500; the most extensive samples of anthrosols are found on the forested bluffs along the main Amazonian rivers, suggestive of the extensive settlement of the floodplains prior to the 16th century, and destroyed by European conquest (Balée 2003: 283; Denevan 2006: 156–158). Anthropologists have also identified various formerly domesticated landscapes (or "cultural forests") in the Amazon by their larger number of planted fruit trees and useful plant species, including palm, bamboo and liana forests. High densities of semi-domesticated species, such as Brazil nut, can also still be identified hundreds of years since the arrival of the Portuguese (Balée 2013).

Notwithstanding the indigenous holocaust that is the history of colonial and postcolonial Brazil, distinctive native agroforestry practices still dot the Amazon, ranging from planting of trees in home gardens and fields to the management of volunteer seedlings of both cultivated and wild species. As scientific knowledge developed over millennia, they represent technologies that co-evolved with the domestication of native forest species and food production systems (Miller and Nair 2006; Castro 2013: 35). Darrell A. Posey's ethnobotanical studies, for example, have underscored the Kayapó's specialized knowledge of soils, plants, and animals and their interrelationships. The indigenes not only located villages in intermediate zones between forest and savannah to maximize access to biodiversity (Posey 2002: 6), in the tropical forest patches or "islands" (*apêtê*) created and managed by the Kayapó in the savannah regions, the indigenes planted more than three-quarters of the 120 plant species sampled by Posey, which were used to attract game and birds and for food, water, medicine, firewood, ceremonial items and everyday materials (Posey 2002: 195–196, 204–205). Alongside hunting and foraging trails, researchers also observed 185 planted trees from more than 15 different species and approximately 1,500 medicinal plants and 5,000 food-producing plants (Posey 2002: 207). Environmental resource management was closely linked to gender roles: male hunters managed the plantings along trekking trails or those used to attract game; other plants were associated with specific age cohorts and kin groups, while women who remained in the village oversaw plant breeding, protection, propagation and soil science (Hecht and Posey 1989: 179–180).

Comparable botanical and pharmacological mastery has been documented for other native Amazonian societies. A survey conducted among the Carib-speaking Kuikuru on the upper Xingu river in the late 1970s found that the community made use of 76 percent of the trees, reflective of their extensive ethnobotanical knowledge of local flora. Among the Waimiri-Atroari, 79 percent of the tree species in one hectare of upland forest were assigned utilitarian properties, including use for medicines, dyes, adornments, transport, poisons, weaponry, construction, and fuel and transport (Milliken et al. 1992: 1, 117).

Throughout the Amazon, extensive indigenous vocabularies to distinguish flora and fauna likewise bespeak heightened environmental knowledge. The Kayapó, for example, recognize 56 folk species of bees, reflecting the rich, empirical knowledge of the environments that they inhabit (Posey 2002: 137). The Kayapó also discriminate between eight forest types: true forest, liana forest, high forest, forest in which light penetrates to the ground, forest gaps, gallery forest, high dark forest and forest transition zone (Hecht and Posey 1989: 178). Likewise, in his research on the Ka'apor people, William Balée affirmed the linguistic differentiation by the indigenes of fallow forests (which the indigenes call *taper*) from high forests (known as *ka'a-te*), and the delineation of their respective species. Furthermore, Balée notes that native people's classificatory models typically name wild plants after domesticates, reflecting the ways in which such conceptual dichotomies are cognitively embedded (Balée 2003: 281–282).

Indigenous agroforestry in the Brazilian Amazon stands in marked contrast to the dominant developmentalist model of indiscriminate clearing for monocrop agriculture and cattle ranching, which leaves the soil of forest openings vulnerable to nutrient depletion and compacting from heavy rains (Posey 2002: 198–199). Ethnobotanical knowledge, transmitted inter-generationally for centuries, is threatened not only by land loss and environmental destruction, but through alteration of native forest management methods and attitudes towards plant resources and substitution of industrial goods for botanical materials (Milliken et al. 1992: 46). Given the decline in soil fertility in the contemporary Amazon due to deforestation, the pervasiveness of crop failure and the instability of land tenure, there is much to learn from indigenous forms of sustainable agriculture and forest management (Clement 2006: 38–39).

Indigenous ontologies and the environment

Indigenous understandings of "nature" also reveal the varied ontologies constitutive of human diversity. Many indigenous Amazonian societies, for example, believe that objects and beings, both animate and inanimate, have spirits, mental faculties and consciousness. According to such "animist" cosmologies, spirits are not exclusive to the human realm, but rather a shared natural (although non-identical) attribute of all beings. In his research on the Achuar (on the Amazonian border of Peru and Ecuador), Philippe Descola noted adoptive relations of kinship with plants and animals, who were believed to have intelligence, spirit, affect and communicative powers that enable social relations with humans (Descola 1994). Imputing such attributes to natural beings allows for their common humanity, subverting the structuralist notion of a nature-culture divide (Turner 2009: 14–17). Similarly, anthropologist Eduardo Viveiros de Castro notes that distinctions between nature and culture are inapplicable to many Amerindian cosmologies, if nature is understood as universal, objective and physical, and culture as particular, subjective and social. Rather, according to his theory of perspectivism, indigenous Amazonians believe that animals subjectively identify as humans, thereby attesting to a universal spiritual (cultural) identity that links animals and humans as alike and distinguished solely by their external or "multinatural" forms. Viveiros de Castro differentiates between Western thought, which "postulates a physical continuity and a metaphysical discontinuity between humans and animals," and Amerindian cosmologies that uphold the metaphysical continuity of the spirit or soul and the discontinuity of the body (Castro 1998: 479; Turner 2009: 11).

In this vein, anthropologists have noted native Amazonians' conceptualization of history as the product of interactions between different types of persons (human and nonhuman) that reaffirm the importance of the process of transformation (Fausto and Heckenberger 2007: 14). As Terry Turner notes, "in many Amazonian cosmologies, the spirit-forms of things (humans, animals, plants, celestial bodies, spirits of the dead) immanently possess the generative capacity that endow species with their specific characteristics and behaviours, including the ability to produce themselves through transformation" (Turner 2009: 32). Historical agency derives from "the fabrication of people's agential capacities through involvement in shamanic and ritual practices" (Fausto and Heckenberger 2007: 14).

Native Amazonian myths – which are best viewed as alternative or complementary to Western forms of historical consciousness in providing shared interpretive frameworks to render the past meaningful and guide social actions – elucidate such beliefs. (Whitehead 2006: xi; Fausto and Heckenberger 2007: 7–8) Fernando Santos-Granero notes a common Amazonian belief that the mythical times in which all things were human (or appeared to each other as human) came to an end due to ancestral fallibility, engendering the different categories of beings that inhabit the contemporary world. The transformation involved multiple metamorphoses, entailing processes of bodily deconstitution and reconstitution, including the interchanging of body parts and artefacts. In other words, "the coming into being of the present-day world was not the result of a creation ex nihilo, but rather the product of the transformation of existing things." (Santos-Granero 2009: 4) The Arawak-speaking Baniwa people of Northwestern Amazonia, for example, who live primarily from horticulture and fishing, believe that the sacrificial acts of creator deities and other primordial beings resulted in distinct species of cultivated and wild plants as gifts to humankind. Creation stories individuate the histories of tobacco, capsicum pepper and manioc as gifts of the deity's body and knowledge to humanity. Poisonous plants, such as curare, are believed to have been generated from the fluids and bodily parts of primordial beings in revenge for their murder or that of their kin (Wright 2009). So, too, the transformation in the human life cycle transcends the boundaries of species upon death. As Turner notes, for example, the Kayapó believe in the transformation of their deceased kin into spirits, or animal-like beings who seek to induce the living into joining them in the spiritual world. Human culture is understood as the incremental social transformation of natural elements, and its reversion through death, rather than a binary contrast between nature and culture. The Kayapó uphold an analogous view for plants and animals, believed to fuse and transform the form and content of their respective species through the conscious force of their distinctive spirits (Turner 2009: 28–37). The assault on indigenous communities in Brazil resulting from deforestation, modern industrial land, and mass media and consumption use not only endangers biological diversity but the cultural patrimony of humankind.

Indigenous ecologies and contemporary conflicts

After centuries of pandemic, enslavement and dispossession, Brazil's rebounding indigenous populations face the furies of 21st-century environmental destruction. Although European colonialism was long intertwined with land-use change, clearly "something new under the sun" marked 20th-century global environmental history, as historian John McNeill has argued, and Brazil's native populations know all too well. Over the course of the 20th century, the world's population quadrupled from 1.5 billion to 6 billion, its urban population increased by 13 times, and its economy grew 15-fold – albeit marked by vast global inequalities. The total increase in industrial output increase 40-fold and the world's total energy use increased 13-fold.

Agro-industrial and urban expansion triggered an endless search for coal, oil, gas, water, timber, mineral ores and foodstuffs. Renewable resources such as forests and woodlands came under particular strain, shrinking by one fifth over the course of the century; about half of this deforestation took place in the tropics after the 1960s, primarily to enable farming and ranching, as well as for timber (McNeill 2000; McNeill 2003). Alongside such change came greater state and nongovernmental concern with environmental protection: at the time of the U.N. Conference on the Human Environment in Stockholm in 1972, few state environmental bureaucracies existed worldwide. When Brazil created the Special Secretariat of the Environmental in 1973, only seven countries had a similar agency (Hochstetler and Keck 2007: 27). By 1988, approximately 68 had been created, and a decade later nearly every nation had a government bureau dedicated to environmental affairs (Khagram 2004: 16). Between 1985 and 1990, membership in the Environmental Defense Fund, the National Resources Defense Council, Greenpeace and the Nature Conservancy doubled, while World Wildlife Fund-US quintupled. Transnational environmental networks increased from two groups in 1953 to 90 in 1993, or from 1.8 percent of total international NGOS to 14.3 percent (Keck and Sikkink 1998: 10–11, 128).

As a rapidly industrializing and resource-rich nation, Brazil's 20th-century environmental history has paralleled and propelled global environmental change. From a population of 17.4 million in 1900 that was 90 percent rural, roughly one century later Brazil grew to a population of 180 million that was 82 percent urban. As Brazilian industry expanded – reaching 37.5% of GDP by the early 2000s – electricity and petroleum came to substitute firewood and charcoal as energy sources by mid-century (Oliveira et al. 2011: 51–64). Between the 1950s and 2000, Brazil also built approximately 300 dams for generating hydroelectric power and flood control, ranking among the top ten nations in terms of large dam construction in the 20th century (Khagram 2004: 144–145). Since the 1970s, 100 million Brazilians gained access to adequately treated water (although nearly 40 million are still deprived), while in the last 30 years, access to electricity increased from 490 kilowatt-hours per capita to about 2,300 kilowatt-hours per capita, largely through the expansion of hydropower (Thomas 2006: 55). Increased air and water pollution in Brazil's large metropolises, coincident with the emergence of new social movements and re-democratization in the late 1980s, engendered greater support for environmental protection among the nation's middle class (Hochstetler and Keck 2007).

Rapid urban growth stimulated and was sustained by rural modernization and frontier development. Between 1945 and 1970, Brazilian agriculture expanded through the incorporation of over 60 million hectares of "new" lands into farms, which often entailed the dispossession of native and traditional peoples (Baer 2001: 400). Today, Brazil is the world's second largest agroexporter: its leading crops include coffee, soy, maize, sugar, oranges and manioc. Agriculture and agroindustries accounted for 27 percent of the GDP in the early 2000s, employing 23 percent and ten to 15 percent of the workforce, respectively (Thomas 2006: 84–85). In 2010, agro-exports reached US$76 billion and represented 38 percent of total exports; soybeans, meat and sugar-ethanol complex products, as well as the forestry sector, are highly represented (Martha Jr., Contini, and Alves 2012: 208). Brazil is also the world's leading beef exporter, with the largest cattle herd in the world, at nearly 200 million head of cattle (Thomas 2006: 76).

The Amazon region, the world's largest rainforest holding a fifth of the world's fresh water and a quarter of its known biodiversity (including 40,000 different plants, 3,000 fish species, 1,300 hundred birds and 1,200 mammals, reptiles and amphibians) has particularly been affected (Greenpeace 2016b). Unlike the Atlantic forests, which suffered sustained (and intensified) effects of Luso-Brazilian economic development over centuries, only one percent of Amazonian forest had been destroyed by the mid-20th century. The lag in deforestation owed to numerous factors: the region's peripheral position in the slave-plantation complexes of the Atlantic world; the comparatively limited impact of the turn of the 20th-century rubber boom on deforestation; and the epidemiological and demographic challenges that Luso-Brazilian statesmen and settlers faced in modifying land use strategies in the region that resulted, in part, from the collapse and reconfiguration of native subsistence and extractive economies under colonial and 19th-century rule (Weinstein 1983; Pádua 2010: 130–131).

During the 1970s and 1980s, in the name of regional integration and national security, Brazil's military government pursued aggressive development of the Amazon through road construction, subsidized agribusiness and directed colonization. The military government offered over $3 billion in subsidies to cattle ranches in the Amazon, predominantly medium- and large-scale properties whose owners, including multinational corporations, received generous tax breaks (London and Kelly 2007: 147–148). Since the colonial period, cattle have served as an advance guard for frontier settlement, ecosystem and land-use modification, and displacement of indigenous peoples in the Americas (Crosby 1972).

In subsequent decades, the military's efforts in the Amazon bore fruit: from 1990 to 2005, the cattle population in the Amazon increased from 26.2 million to 65 million, largely serving the urban domestic market (London and Kelly 2007: 147–148). Consequently, over the last half century, the Amazon forest was reduced by 15 percent – mostly concentrated in the states of Rondônia, Pará and Mato Grosso – due to cattle ranching, agriculture and logging (Thomas 2006: 73–75). Eight-five percent of deforestation has occurred near roads, whose network doubled by over 50,000 miles from the 1970s to 2000; and 75 percent of deforested land is occupied by medium and large cattle ranches (Thomas 2006: 76; London and Kelly 2007: 150). The Amazon's population now numbers 25 million people, representing Brazil's fastest growing region between 2000 and 2010 due to migration to the agricultural frontier (Romero 2012).

As in urban centres, Brazil's countryside is mottled by stark disparities in wealth and power. Alongside a highly capitalized agricultural sector, millions struggle with reduced landholdings, degraded natural resources, and inadequate infrastructure and social services (Thomas 2006: 86). The environmental costs of development reflect such disparity. The top decile of Brazil's population, enjoying vast access to goods and services, generate a disproportionate amount of waste and pollution, while millions lack access to decent health and sanitation services, and cling to environmentally precarious areas (Baer 2001: 396). Since natural resources comprise a larger proportion of assets for Brazil's poor rural populations – including its indigenous peoples – than for its privileged social sectors, environmental conflicts amount to basic fights for survival (Thomas 2006: 70). Economist Joan Martinez-Alier and historian Ramachandra Guha have deemed the latter's concomitant ecological consciousness and praxis the "environmentalism of the poor" (Guha and Martinez-Alier 1997).

Unlike the "cult of wilderness" enshrining nature's transcendental value and its requisite preservation through the creation of national parks, or the "gospel of eco-efficiency" touting the importance of natural resource conservation for long-term development, the environmentalism of the poor privileges the importance of the natural environment as the basis of livelihood and social justice (Martinez-Alier 2005; Guha 2000: 104–105). Although indigenous ties to the environment, as previously noted, encompass distinct cosmological, symbolic and historical realms of meaning-making that are not reducible to cultural materialism, this current of environmental thought effectively advocates for the sustainable coevolution of traditional peoples and landscapes threatened by agribusiness, mining and large-scale public infrastructure projects (Martinez-Alier 2005: 1–15). The concluding section offers three case studies that shed light on contemporary environmental challenges faced by Brazil's native peoples.

Cattle, soybeans, sugar and dams

Prior to the mid-1900s, the Xavante, a Gê-speaking people of Mato Grosso, had enjoyed dominion over vast stretches of *cerrado*, or tropical savannah, for more than a century. As hunters and gatherers who spent most of the year on trek, the Xavante possessed intimate knowledge of the region's flora and fauna. Time-tested practices of fire management were applied to hunting game and agriculture, while native bellicosity towards outsiders ensured territorial defence. With the mid-century Brazilian state spearheading the settlement and integration of eastern Mato Grosso, the Xavante pursued sustained peaceful contact with the dominant society between the mid-1940s and early 1960s (Garfield 2001). Numbering today some 18,000 (one of Brazil's largest indigenous groups) and dispersed among nine reservations, the Xavante face severe problems due to the drastic transformations that have engulfed their communities over the previous half century.

Since the 1970s, more than one million square kilometres of cerrado (an area equivalent to half the size of Mexico) were turned into pastureland or commercial agricultural production (Thomas 2006: 80). Key to the model of agricultural modernization was subsidized credit for capital financing, rural extension and support for agricultural research focusing on productivity gains. Entrusted with agricultural research and development, EMBRAPA, the Brazilian corporation created in 1973, strategically coordinated initiatives with multiple state agricultural research stations, universities and agricultural colleges, and private partners in Brazil and abroad. The agency focused heavily on the cerrado region, comprising 24 percent of national territory, whose acidic, low-fertility soils required the introduction of Green Revolution technologies, including cultivation of high-yield grains and cereals, intensive use of fertilizers and heavy machinery, and vast use of pesticides and drugs for crop cultivation and livestock production (Martha, Contini, and Alves 2012: 402–404).

The Xavante recoiled as the world around them collapsed. In Mato Grosso, the cattle herd grew from 6.5 million head in 1985 to 26 million head in 2008. Soy cultivation increased from 56,000 hectares to 4.5 million hectares during roughly the same period, reaching nearly one-third of national production of 62 million tons, or 28 percent of the world's total, earning $10 billion from exports (London and Kelly 2007: 171; Mueller 2012: 191). Increased soybean production occurred primarily through expanded cultivation rather than higher yield per acre; as a legume, fixing atmospheric nitrogen in the soil, soy is not as fertilizer-responsive as corn and other grains. In Mato Grosso, clearing of the wooded grasslands and gallery forests of the cerrado paved the way for new acreage for soy (Brown 2012: 93–100). Demand was driven largely by European and Chinese consumption: the appearance of mad cow disease in Europe in 1986 elevated soy in the diet of the continent's livestock, while the expanding Chinese middle-class appetite for meat led to a tenfold increase in soy imports for animal feed between 1999 and 2003 (London and Kelly 2007: 169).

Commercial agriculture and ranching contributed to a startling loss of biodiversity, depletion of animal game and modification of soils in the cerrado which have irrevocably impacted the traditional livelihoods of indigenous peoples (Baer 2001: 402–404.) Crowded onto reservations largely created in the 1980s, expanding Xavante communities contend with reduced prospects for hunting, degraded soils for agriculture and alarming rates of hypertension, diabetes and obesity due to increasingly sedentary lifestyles and overconsumption of sugary, high-fat processed foods. They also face perennial scorn from pro-business advocates covetous of their territories and/or disdainful of their alternative lifestyles. Collaboration with environmental non-governmental organizations has allowed some Xavante communities to undertake projects aimed at replenishing or managing endangered animal game and valorizing traditional botanical knowledge (Graham 2009).

Agribusiness also threatens the survival of the Guarani Indians, once numbering more than a million and reduced to some 51,000 people after centuries of contact with Europeans and their descendants. Divided into three groups – the Kaiowá (with an estimated population of 31,000 as of 2008), the Ñandeva (1,300) and the Mbya (7,000) – Brazil's Guarani population is located principally in the state of Mato Grosso do Sul. Corralled onto reservations devised by state bureaucracies during the first part of the 20th century, the Guarani contend with territorial boundaries reflective of public policies promoting land commodification and indigenous integration (through wage labour and co-optation of native leadership) rather than the defense of traditional native homelands (Instituto Socioambiental 2011).

For the Guarani, land has been the structure of life – the source of not only their physical survival, but their cultural reproduction and spiritual well-being. Guarani cosmographies hallow the notion of *tekoha*, the physical environment comprised of land, brush, water, flora and fauna that sustain the various extended family groups that occupy a determined physical area. Ideally, it includes *ka'aguy* (scrubland) for cultivating small plots and raising livestock, foraging, access to timber, water and medicinal plants. But the ka'aguy is also the cosmological hearth of spiritual beings and the stage for myth narration. As such, tekoha encompasses the ecological, social and religious realms, and its impairment, attenuated by diminished territory and physical and cultural violence, endangers the sociocultural reproduction of Guarani communities. The ideal of reclaiming cultural integrity through restoration of tekoha and ancestrally sacred territories drives contemporary land mobilization strategies and represents a desperate fight for ethnic survival (Instituto Socioambiental 2011)

On the Dourados reserve, for example, 12,000 Guarani are crammed into a territory of roughly 7,000 acres. Local agricultural production fails to sustain indigenous nutritional balance and consumer needs, while surrounding cattle ranches and farms have depleted or destroyed animal game and fish sources. Since the 1980s, sugar plantations in Mato Grosso do Sul have benefited from state promotion of a domestic ethanol industry to offset Brazil's historic dependency on petroleum imports and growing international consumption. Nationally, sugarcane production increased 369 percent between 1976 and 2011, growing from 6.72 million tons to 31.51 million tons, with Brazilian law requiring a minimum blend of ethanol in domestic gasoline to protect sugar growers (Martha Jr., Contini, and Alves 2012: 207). By 2015, sugar cane accounted for 16.9% of Brazil's energy needs – the largest source of renewable energy – and a quarter of global ethanol use. With the expansion of agribusiness and biofuels, the Kaiowá have become increasingly besieged. The dependency of indigenous men on seasonal employment at regional sugar mills and plantations has compounded interethnic tensions (Survival International 2010).

During the last three decades, Guarani-Kaiowá communities in Mato Grosso do Sul have been ravaged by mass suicide. Over 600 Guarani, mainly young adults and children, have taken their lives in the face of territorial dispossession, poverty, hunger, cultural violence and racial discrimination. The suicide rate among the Guarani of Mato Grosso do Sul was 19 times national indices in the early 2000s. The Guarani infant mortality rate is also more than double the national average, while life expectancy is more than 20 years lower than the national average: only 7.8 percent of the Dourados reserve's population is older than 50 (Diário Mato Grosso do Sul 2012).

Through persistent litigation and land "occupations," the Guarani had recouped 16 tekoha by 2003, but ranchers and farmers have also repelled native efforts with systematic violence (Instituto Socioambiental 2011; United Nations 2016). In 2014, Mato Grosso do Sul witnessed the highest number of murders of indigenous peoples in Brazil. Bemoaning her people's living conditions, Marta Silva affirmed: "We Indians don't want money or riches. Do you know what we want? We just want enough land to live on how we like" (Survival International 2000: 64).

The indigenous opponents of the São Luiz do Tapajós hydroelectric dam, whose manifesto serves as the prologue to this essay, have confronted yet another peril. In the early 2000s, 90 percent of Brazil's electrical energy consumption derived from hydropower (Thomas 2006: 9). The São Luiz do Tapajós, one of 40 medium-sized to large dams that the Brazilian government planned in the biodiverse Amazon region as a whole, threatened to flood 400 square kilometres of rainforest and contribute to an additional 2,200 square miles of deforestation. For indigenous populations of the basin, the dam would trigger mass deaths of fish and reduce floodplain fertility due to interrupted sediment flows. Government officials initially disregarded ancestral claims of the Munduruku, an indigenous group of 12,000 spread across 128 villages and most potentially affected by the dam, in refusing to demarcate indigenous territory (Greenpeace 2016b).

For indigenous communities of the Tapajós basin, the historical record was foreboding. Under the military government, Brazil constructed a number of large-scale dams, with the World Bank investing more than $3 billion in the nation's energy sector and the Inter-American Development Bank providing Brazil with another $2 billion. Politicians, engineers, industrialists, bureaucrats and multinational corporations that supported large dam construction in Brazil heralded the increased potential for energy supply (in the face of higher prices for foreign petroleum), irrigation and flood control. As the quintessential subjugation of nature by science and technology, dams also epitomized the ethos of national progress in late developing countries (Khagram 2004: 1–10, 144–145). Local communities were traditionally excluded from the planning and implementation of large dam projects in Brazil and received inadequate compensation for territorial displacement and resettlement.

The massive Itaipu dam, completed in 1982 on the Paraná river in southern Brazil by a joint Brazilian-Paraguayan public enterprise with a generating capacity of 14,000 megawatts (and cost overruns of $3.3 billion), for example, inundated over one million square kilometres of land and displaced more than 42,000 people, including the Avá-Guarani people. In the Amazon region, the 8,000-megawatt Tucuri Dam on the Tocantins river, completed in 1984 with the objective of providing energy to large aluminium-processing plants, had flooded over 2,400 square kilometres of the Parakanã Indian reservation. The Parakanã, contacted by Brazil's government in 1971 to advance the construction of the Trans-Amazon highway, had already been relocated by the Indian agency five times during the 1970s and suffered catastrophic loss due to epidemics. Those displaced by the Tucuruí dam suffered high rates of mortality due to malaria and other diseases (Davis 1977: 66–68; Fearnside 1999). Similarly, the 250-megawatt Balbina dam on the Uatumã river, initiated in 1973 with the objective of providing the Amazonian city of Manaus with electric energy, flooded two Waimiri-Atroari villages and displaced one-third of the native population. Balbina failed to achieve even one quarter of its energy-generating potential (Gomes 2000: 195–196; Khagram 2004: 144–145).

In the mid-1980s, anti-dam actors in Brazil, linked to transnational environmental networks, had successfully blocked efforts by state electrical company Eletronorte to build 50 dams along the Amazon's Xingu river, pressuring the World Bank to withhold funding (Khagram 2004: 155). And Brazil's Constitution of 1988, promulgated after two decades of authoritarian rule, mandated approval of mining and hydroelectric projects on indigenous lands by the national congress, and prior and informed consent by indigenous peoples on matters that affect their communities. The International Labour Organization Indigenous and Tribal Peoples Convention 169 of 1989, ratified in Brazil in 2002 and enacted by presidential decree in 2004, grants indigenous peoples self-management and self-regulation of their territories and self-determination and autonomy (United Nations 2016). The Convention applies to

> peoples in independent countries who are regarded as indigenous on account of their descent from the populations which inhabited the country, or a geographical region to which the country belongs, at the time of conquest or colonization or the establishment of present state boundaries and who, irrespective of their legal status, retain some or all of their own social, economic, cultural and political institutions

and to

> tribal peoples in independent countries whose social, cultural and economic conditions distinguish them from other sections of the national community, and whose status is regulated wholly or partially by their own customs or traditions or by special laws or regulations.
>
> *(International Labour Organization 1989)*

In 2001, however, after Brazil experienced highly unpopular electricity rationing due to drought-related reduction in hydropower, the federal government relaunched military-era plans for the construction of a series of hydroelectric dams in the Amazon. The Brazilian government justified the projects in the name of national security – the Constitution allows for the removal of indigenous peoples from their lands in case of natural disaster or national security pending authorization by the national congress – and as a clean solution to the nation's energy needs. In fact, dams contribute to the greenhouse effect through the release of methane, sulphur and other gases deriving from the decomposition of submerged flora (Greenpeace 2016b).

In 2011, the Brazilian government approved the construction of the Belo Monte dam on the Xingu river with funding provided by Brazil's National Development Bank. The dam would serve to furnish energy for large mining and construction company operations in the Amazon and to provide electric power for São Paulo and Rio de Janeiro. When fully operational, the dam's reservoir is expected to flood more than 500 square kilometres of land, divert the flow of the Xingu, and displace between 20,000 and 40,000 people from their homes, including over 800 indigenous people. Food and water supplies for local populations will also be endangered (Diamond and Poirier 2010; Amazon Watch 2011). Although Brazil's Indian agency and public prosecutor objected to granting an operational license to Norte Energia S.A., the consortium that would build Belo Monte, on the grounds that the project failed to mitigate social and environmental impacts, the license was granted. The U.N. Special Rapporteur on the Rights of Indigenous Peoples and the Inter-American Commission on Human Rights denounced the Belo Monte dam for violating indigenous peoples' rights, and native indigenous groups in the basin, including the Kayapó, Juruna and Arara, opposed the project as well. As José Carlos Arara decried, "Our ancestors are there inside this land, our blood is inside the land, and we have to pass on this land with the story of our ancestors to our children" (Diamond and Poirier 2010: 29). Since environmental assessment impacts only consider "directly affected areas," the dam's supporters argued that few indigenous people would suffer the direct consequences of flooding. The reduced water flow, interruption in river navigation, loss of fish species and potential increase in malaria that will inevitably impact indigenous communities were conveniently ignored (Amazon Watch 2011). Indeed, an investigation in February 2016, after the reservoir had begun to be filled, found 16 tons of dead fish dumped by Norte Energia in a landfill. Population influx into the region due to dam construction also increased deforestation and illegal logging, as well as resource depletion on Arara land. Environmental organizations widely consider the $815 million in mitigation funds inadequate to compensate for the grave social and humanitarian costs of the Belo Monte project (Amazon Watch 2011).

The peoples of the Tapajós basin have been temporarily spared such calamity. In April 2016, Brazil's environmental agency suspended the licensing process due to its concerns regarding the SLT's impact on indigenous peoples. The soaring cost overruns and graft surrounding the Belo Monte project, the high operating costs and reduced profits, seasonal fluctuations in energy-producing capacity, and overall downturn of Brazil's economy in recent years undoubtedly gave government officials and corporate investors pause (*The Guardian* 2016). Support from domestic and international non-governmental environmental and indigenous advocacy groups, such as Greenpeace, Amazon Watch and Survival International, provided Brazil's native peoples with important allies in their struggle.

'The Indians: A People Doomed to Die'?

In 1973, dismayed by the military government's aggressive developmental policies in the Amazon region, Catholic bishops in Brazil issued the document *Y Juca-Pirama. The Indians: A People Doomed to Die*. As accusations of genocide blanketed the international press, indigenous peoples, anthropologists, environmentalists, church officials and legal scholars mobilized across Brazil and worldwide to denounce the onslaught on native communities (Davis 1977: 158). Their activism was key to dooming the bishops' prophecy: Brazil's indigenous populations boomed and many of their traditional territories have been officially demarcated. Nevertheless, unyielding global demand for natural resources continues to jeopardize the physical and cultural integrity of Brazil's native communities. Indigenous constitutional rights remain compromised by developmental ideologies, plutocratic interests, government inaction, and systemic forms of racism and discrimination. In the balance lie age-old and adaptive forms of environmental knowledge and management, cosmological realms and ontological beliefs that give life to indigenous communities and cultural diversity to humanity.

Bibliography

Allen, Elizabeth. 2006. 'Fire in Roraima, 1998–Politics and Human Impact: What Role for Indigenous People in Brazilian Amazonia?,' in Darrell Addison Posey and Michael J. Balick (eds.), *Human Impacts on Amazonia: The Role of Traditional Ecological Knowledge in Conservation and Development*, pp. 51–84. New York: Columbia University Press.

Amazon Watch. 2011. *Belo Monte Facts*. http://amazonwatch.org/work/belo-monte-facts (accessed on 7 December 2016).

Argawal, Arun. 1995. 'Dismantling the Divide Between Indigenous and Scientific Language,' *Development and Change* 26 (3): 413–439.

Argawal, Arun and Clark C. Gibson. 1999. 'Enchantment and Disenchantment: The Role of Community in Natural Resource Conservation,' *World Development* 27 (4): 629–649.

Baer, Werner. 2001. *The Brazilian Economy: Growth and Development*. 5th edition. Westport, CT: Praeger.

Balée, William. 2003. 'Native Views of the Environment in Amazonia,' in Helaine Selin (ed.), *Nature across Cultures: Views of Nature and the Environment in Non-Western Cultures*, pp. 277–288. Dordrecht; Boston: Kluwer Academic Publishers.

Balée, William. 2013. *Cultural Forests of the Amazon: A Historical Ecology of People and Their Landscapes*. Tuscaloosa, AL: The University of Alabama Press.

Brazil's Diverse Energy Matrix. 2016. http://sugarcane.org/the-brazilian-experience/brazils-diverse-energy-matrix (accessed on 7 January 2017).

Brown, Lester R. 2012. *Full Planet, Empty Plates: The New Geopolitics of Food Scarcity*. New York: W.W. Norton.

Castro, Eduardo Viveiros de. 1998. 'Cosmological Deixis and Amerindian Perspectivism,' *The Journal of the Royal Anthropological Institute* 4 (3): 469–488.

Castro, Eduardo Viveiros de. 2013. 'Economic Development and Cosmopolitical Re-Involvement: From Necessity to Sufficiency,' in Lesley Green (ed.), *Contested Ecologies: Dialogues in the South on Nature and Knowledge*. Cape Town: HSRC Press.

Clement, Charles R. 1999a. '1492 and the Loss of Amazonian Crop Genetic Resources: I. The Relation between Domestication and Human Population Decline,' *Economic Botany* 53 (2): 188–202.

Clement, Charles R. 1999b. '1492 and the Loss of Amazonian Crop Genetic Resources: II. Crop Biogeography at Contact,' *Economic Botany* 53 (2): 203–216.

Clement, Charles R. 2006. 'Demand for Two Classes of Traditional Agroecological Knowledge in Modern Amazonia,' in Darrell Addison Posey and Michael J. Balick (eds.), *Human Impacts on Amazonia: The Role of Traditional Ecological Knowledge in Conservation and Development*, pp. 33–50. New York: Columbia University Press.

Conklin, Beth and Laura R. Graham. 1995. 'The Shifting Middle Ground: Amazonian Indians and Eco-Politics,' *American Anthropologist* 97 (4): 695–710.

Conselho Indigenista Missionário. 2015. *Violence against Indigenous Peoples in Brazil: 2014 Data*. Brasília: CIMI.

Crosby, Alfred W. 1972. *The Columbian Exchange: Biological and Cultural Consequences of 1492*. Westport, CT: Greenwood.

Crosby, Alfred W. 1986. *Ecological Imperialism: The Biological Expansion of Europe, 900–1900*. Cambridge; New York: Cambridge University Press.

Davis, Shelton H. 1977. *Victims of the Miracle: Development and the Indians of Brazil*. Cambridge; New York: Cambridge University Press.

Dean, Warren. 1995. *With Broadax and Firebrand: The Destruction of the Brazilian Atlantic Coastal Forest*. Berkeley: University of California Press.

Denevan, William M. 1992. 'The Pristine Myth: The Landscape of the Americas in 1492,' *Annals of the Association of American Geographers* 82 (3): 369–385.

Denevan, William M. 2003. *Cultivated Landscapes of Native Amazonia and the Andes*. Cambridge: Cambridge University Press.

Denevan, William M. 2006. 'Pre-European Forest Cultivation in the Amazon,' in William Balée and Clark L. Erickson (eds.), *Time and Complexity in Historical Ecology: Studies in the Neotropical Lowlands*, pp. 153–163. New York: Columbia University Press.

Descola, Philippe. 1994. *In the Society of Nature: A Native Ecology in Amazonia*, transl. Nora Scott. Cambridge; New York: Cambridge University of Press.

Descola, Philippe. 2013. *The Ecology of Others*, transl. Geneviève Godbout and Benjamin P. Luley. Chicago: Prickly Paradigm Press.

Diamond, Sara and Christian Poirier. 2010. 'Brazil's Native Peoples and the Belo Monte Dam: A Case Study,' *NACLA Report on the Americas*: 25–29.

Diário Mato Grosso do Sul. 2012. Friday, 14 September. *Apenas 7.8% dos índios de Dourados chegam aos 50 anos, aponta IBGE*. http://diarioms.com.br/apenas-78-dos-indios-de-dourados-chegam-aos-50-anos-aponta-ibge/ (accessed on 16 January 2017).

Diegues, Antônio Carlos Sant'Ana. 2004. *O mito moderno da natureza intocada*. 5th edition. São Paulo: Editora Hucitec.

Dove, Michael R. et al. 2003. 'The Global Mobilization of Environmental Concepts: Re-Thinking the Western/ Non-Western Divide,' in Helaine Selin (ed.), *Nature across Cultures: Views of Nature and the Environment in Non-Western Cultures*, pp. 19–46. Dordrecht; Boston: Kluwer Academic Publishers.

Fausto, Carlos and Michael Heckenberger. 2007. 'Indigenous History and the History of the "Indians",' in Carlos Fausto and Michael Heckenberger (eds.), *Time and Memory in Indigenous Amazonia: Anthropological Perspectives*, pp. 1–43. Gainesville: University Press of Florida.

Fearnside, Philip M. 1999. 'Social Impacts of Brazil's Tucuruí Dam,' *Environmental Management* 24 (4): 483–495.

Garfield, Seth W. 2001. *Indigenous Struggle at the Heart of Brazil: State Policy, Frontier Expansion, and the Xavante Indians.* Durham, NC: Duke University Press.

Gomes, Mercio P. 2000. *The Indians and Brazil,* transl. John W. Moon. Gainesville: University Press of Florida.

Graham, Laura R. 2009. 'The Tractor Invasion,' *Cultural Survival Quarterly* 33 (2). www.culturalsurvival.org/publications/cultural-survival-quarterly/tractor-invasion (accessed on 4 January 2017).

Greenpeace. 2016a. *Brazil: The Most Dangerous Country for Environmental Activists in 2015.* www.greenpeace.org/international/en/news/Blogs/makingwaves/brazil-most-dangerous-country-for-environmental-activists-especially-amazon/blog/56871/ (accessed on 26 October 2016).

Greenpeace. 2016b. *Damning the Amazon: The Risky Business of Hydropower in the Amazon.* www.greenpeace.org/international/Global/brasil/documentos/2016/Greenpeace_Damning_The_Amazon-The_Risky_Business_Of_Hydropower_In_The_Amazon-2016.pdf (accessed on 31 October 2016).

The Guardian. 2016. Friday, 22 April. 'Brazil Amazon Dam Project Suspended Over Concerns for Indigenous People.' www.theguardian.com/world/2016/apr/22/brazil-amazon-dam-project-suspended-indigenous-munduruku-sao-luiz-do-tapajos (accessed on 10 December 2016).

Guha, Ramachandra. 2000. *Environmentalism: A Global History.* New York: Longman.

Guha, Ramachandra and Joan Martinez-Alier. 1997. *Varieties of Environmentalism: Essays North and South.* London: Earthscan Publications.

Hecht, Susanna B. and Darrell A. Posey. 1989. 'Preliminary Results on Soil Management Techniques of the Kayapó Indians,' *Advances in Economic Botany* 7: 174–188.

Hochstetler, Kathryn and Margaret E. Keck. 2007. *Greening Brazil: Environmental Activism in State and Society.* Durham, NC: Duke University Press.

Instituto Socioambiental. 2011. *Povos Indígenas no Brasil.* https://pib.socioambiental.org/pt/povo/guarani-kaiowa/560 (accessed on 20 December 2016).

Inter-American Commission on Human Rights. 1997. *Report on the Situation of Human Rights in Brazil.* www.cidh.org/countryrep/brazil-eng/chaper%206%20.htm (accessed on 18 December 2016).

International Labour Organization. 1989. *Indigenous and Tribal Peoples Convention, 1989 (No. 169).* www.ilo.org/dyn/normlex/en/f?p=NORMLEXPUB:12100:0::NO::P12100_INSTRUMENT_ID:312314. (accessed on 15 December 2016).

Jones, Patrice M. 2000. Thursday, April 20. 'Alcoholism, Malnutrition Ravage Brazil Tribes,' *Chicago Tribune.* http://articles.chicagotribune.com/2000-04-20/news/0004200144_1_brazilian-indians-indian-village-struggle-for-indian-rights (accessed on 27 December 2016).

Keck, Margaret E. and Kathryn Sikkink. 1998. *Activists beyond Borders: Advocacy Networks in International Politics.* Ithaca, NY: Cornell University Press.

Khagram, Sanjeev. 2004. *Dams and Development: Transnational Struggles for Water and Power.* Ithaca, NY: Cornell University Press.

Latour, Bruno. 1993. *We Have Never Been Modern,* transl. Catherine Porter. Cambridge, MA: Harvard University Press.

Lima, Deborah and Jorge Pozzobon. 2005.'Amazônia socioambiental. Sustentabilidade ecológica e diversidade social,' *Estudos Avançados* 19 (54): 45–76.

Little, Paul E. 2001. *Amazonia: Territorial Struggles on Perennial Frontiers.* Baltimore: Johns Hopkins University Press.

London, Mark and Brian Kelly. 2007. *The Last Forest: The Amazon in the Age of Globalization*. New York: Random House.

'Manifesto of the Alliance among the Kayabi, Apiaká, Munduruku and Rikbaktsa Indigenous Peoples.' 2015. http://amazonwatch.org/assets/files/2015-04-24-teles-pires-manifesto.pdf (accessed on 3 November 2016).

Mann, Charles C. 2005. *1491: New Revelations of the Americas before Columbus*. New York: Knopf.

Martha Jr., Geraldo B., Elisio Contini and Eliseu Alves. 2012. 'Embrapa: Its Origins and Changes,' in Werner Baer (ed.), *The Regional Impact of National Policies: The Case of Brazil*, pp. 204–226. Cheltenham: Edward Elgar.

Martinez-Alier, Joan. 2005. *The Environmentalism of the Poor: A Study of Ecological Conflicts and Valuation*. New Delhi; New York: Oxford University Press.

McNeill, John R. 2000. *Something New under the Sun: An Environmental History of the Twentieth-Century World*. New York: W.W. Norton.

McNeill, John R. 2003. 'Resource Exploitation and Over-Exploitation: A Look at the 20th Century,' in Brigitta Benzing and Bernd Herrmann (eds.), *Exploitation and Over-exploitation in Societies Past and Present*, pp. 51–60. Münster; London: LIT.

Miller, R.P. and P.K.R. Nair. 2006. 'Indigenous Agroforestry Systems in Amazonia: From Prehistory to Today,' *Agroforestry Systems* 66 (2): 151–164.

Milliken, William et al. 1992. *The Ethnobotany of the Waimiri Atroari Indians of Brazil*. London: Royal Botanic Gardens, Kew.

Moore, Jason W. 2015. *Capitalism in the Web of Life: Ecology and the Accumulation of Capital*. New York: Verso.

Moran, Emilio. 1993. *Through Amazonian Eyes: The Ecology of Amazonian Populations*. Iowa City: University of Iowa Press.

Moreira Neto, Carlos de Araújo. 1988. *Índios da Amazônia: de maioria a minoria (1750–1850)*. Petrópolis: Vozes.

Mueller, Charles C. 2012. 'Regional Development and Agricultural Expansion in Brazil's Legal Amazon: The Case of the Mato Grosso Frontier,' in Werner Baer (ed.), *The Regional Impact of National Policies: The Case of Brazil*, pp. 184–203. Cheltenham: Edward Elgar.

Nixon, Rob. 2011. *Slow Violence and the Environmentalism of the Poor*. Cambridge, MA: Harvard University Press.

Oliveira, Adilson de et al. 2011. 'Energy Restrictions to Growth: The Past, Present and Future of Energy Supply in Brazil,' in Edmund Amann, Werner Baer, and Donald V. Coes (eds.), *Energy, Bio Fuels and Development: Comparing Brazil and the United States*, pp. 51–64. London; New York: Routledge.

Oliveira Filho, João Pacheco de, ed. 1999. *A Viagem da volta: etnicidade, política e reelaboração cultural no nordeste indígena*. Rio de Janeiro: Contra Capa.

Pádua, José Augusto. 2010. 'European Colonialism and Tropical Forest Destruction in Brazil: Environment beyond Economic History,' in John R. McNeill, José Augusto Pádua, Manesh Rangarajan (eds.), *Environmental History: As If Nature Existed*. Oxford; New York: Oxford University Press.

Portal Brasil. 2015. *No Brasil, população indígena é de 896,9 mil*. www.brasil.gov.br/governo/2015/04/populacao-indigena-no-brasil-e-de-896-9-mil (accessed on 17 November 2016).

Posey, Darrell Addison. 2002. *Kayapó Ethnoecology and Culture*, ed. Kristina Plenderleith. London; New York: Routledge.

Ramos, Alcida Rita. 1998. *Indigenism: Ethnic Politics in Brazil*. Madison: University of Wisconsin Press.

Ribeiro, Darcy. 1970. *Os índios e a civilização: a integração das populações indígenas no Brasil moderno*. Rio de Janeiro: Civilização Brasileira.

Romero, Simon. 2010. Saturday, 24 November. 'Swallowing Rainforest, Cities Surge in Amazon,' *New York Times*. www.nytimes.com/2012/11/25/world/americas/swallowing-rain-forest-brazilian-cities-surge-in-amazon.html (accessed on 13 October 2016).

Santos-Granero, Fernando. 2009. 'Introduction: Amerindian Constructional Views of the World,' in Fernando Santos-Granero (ed.), *The Occult Life of Things: Native Amazonian Theories of Materiality and Personhood*, pp. 1–29. Tucson: University of Arizona Press.

Sponsel, Leslie. 2011. 'The Religion and Environment Interface: Spiritual Ecology in Ecological Anthropology,' in Helen Kopnina and Eleanor Shoreman-Ouimet (eds.), *Environmental Anthropology Today*. London; New York: Routledge.

Survival International. 2000. *Disinherited: Indians in Brazil*. London: Survival International.

Survival International. 2010. *Report to UN Reveals Shocking Situation of Guarani Tribe in Brazil*. www.survivalinternational.org/news/5671 (accessed on 12 December 2016).

Survival International. 2016. *The Guarani*. www.survivalinternational.org/tribes/guarani (accessed on 15 December 2016).

Turner, Terry S. 2009. 'The Crisis of Late Structuralism. Perspectivism and Animism: Rethinking Culture, Nature, Spirit, and Bodiliness,' *Tipití: Journal of the Society for the Anthropology of Lowland South America* 7 (1): 1–40.

Thomas, Vinod. 2006. *From Inside Brazil: Development in a Land of Contrasts*. Washington, DC: Stanford Economics and Finance; World Bank.

United Nations. 2016. *Report of the Special Rapporteur on the Rights of Indigenous Peoples on her Mission to Brazil*. https://documents-dds-ny.un.org/doc/UNDOC/GEN/G16/174/05/PDF/G1617405.pdf?OpenElement (accessed on 15 December 2016).

Warren, Jonathan W. 2001. *Racial Revolutions: Antiracism and Indian Resurgence in Brazil*. Durham, NC: Duke University Press.

Weinstein, Barbara. 1983. *The Amazon Rubber Boom, 1850–1920*. Stanford: Stanford University Press.

Whitehead, Neil L. 2006. 'Introduction,' in Neil L. Whitehead (ed.), *Histories and Historicities in Amazonia*, pp. vii–xx. Lincoln: University of Nebraska Press.

Wright, Robin M. 2009. 'The Fruit of Knowledge and the Bodies of the Gods: Religious Meanings of Plants among the Baniwa,' *Journal for the Study of Religion, Nature and Culture* 3 (1): 126–153.

Wright, Robin M., Wolfgang Kapfhammer, and Flavio Braune Wiik. 2012. 'The Clash of Cosmographies: Indigenous Societies and Project Collaboration–Three Ethnographic Cases (Kaingang, Sateré-Mawé, Baniwa),' *Vibrant: Virtual Brazilian Anthropology* 9 (1). www.scielo.br/scielo.php?script=sci_arttext&pid=S1809-43412012000100014 (accessed on 18 July 2016).

INDEX

Note: page numbers in *italics* and **bold** refer to figures and tables respectively.

For Product Safety Concerns and Information please contact our EU representative GPSR@taylorandfrancis.com
Taylor & Francis Verlag GmbH, Kaufingerstraße 24, 80331 München, Germany

www.ingramcontent.com/pod-product-compliance
Lightning Source LLC
LaVergne TN
LVHW010629300926
849162LV00001B/97
* 9 7 8 0 3 6 7 2 4 5 1 7 7 *